TRANSFORMER PRACTICE

MANUFACTURE, ASSEMBLING, CONNECTIONS, OPERATION AND TESTING

BY

WILLIAM T. TAYLOR

FELLOW, A.I.E.E., M.I.E.E. AND M.I.M.E., ETC.

SECOND EDITION
ENTIRELY REWRITTEN—ENLARGED AND RESET

McGRAW–HILL BOOK COMPANY, Inc.
239 WEST 39TH STREET, NEW YORK
6 BOUVERIE STREET, LONDON, E. C.
1913

PREFACE TO SECOND EDITION

The writer seeks to keep this book practical and for a reference on all matters connected with the operation of transformers and static induction apparatus. It is particularly intended for those who are operating or constructing plants or transformers and is written with the view of assisting engineers out of certain operating difficulties which they can readily solve when they are short of the right apparatus and need a temporary arrangement such as using certain changes of phases.

New schemes of any kind will be appreciated and will add to the value of this book.

WILLIAM T. TAYLOR.

CHAPLANCA, PERU, SOUTH AMERICA,
August, 1913.

PREFACE TO FIRST EDITION

Although much has been written on the fundamental principles of transformers, little data have been published concerning their connection, installation and operation. It was this lack of easily available information and the widespread desire of operators and engineers in the field to possess such information, that led the author to put in type these notes, which had been written up in the course of a number of years of experience in the field.

A working knowledge of the fundamental principles of electrical engineering is presupposed, and for this reason the treatment does not go into the whys and wherefores very deeply, but simply states the facts in as few words as possible. To aid in understanding quickly the phase relations and relative values of the various e.m.fs. and currents involved in a given system, vector diagrams are given with all diagrams of circuit connections.

W. T. Taylor.

Baramulla, Kashmir, India,
  December, 1908.

CONTENTS

		PAGE
PREFACE		v

CHAPTER
I.	Introduction	1
II.	Simple Transformer Manipulations	22
III.	Two-phase Transformer Connections	30
IV.	Three-phase Transformation System	39
V.	Three-phase Transformer Difficulties	84
VI.	Three-phase Two-phase Systems and Transformation	97
VII.	Six-phase Transformation and Operation	109
VIII.	Methods of Cooling Transformers	117
IX.	Construction, Installation and Operation of Large Transformers	136
X.	Auto Transformers	170
XI.	Constant-current Transformers and Operation	178
XII.	Series Transformers and Their Operation	186
XIII.	Regulators and Compensators	209
XIV.	Transformer Testing in Practice	227
XV.	Transformer Specifications	258
	APPENDIX	270
	INDEX	273

TRANSFORMER PRACTICE

CHAPTER I

INTRODUCTION

Development of Art of Transformer Construction

The development of the alternating-current transformer dates back about 25 years. At that time very little was known regarding design for operation at high voltages, and the engineer of the present day can scarcely realize the difficulties encountered in the construction of the early transformers.

The high-voltage transformer made long-distance transmission work possible, and the increased distances of transmission stimulated the design of large transformers. In the early days of transformer development as many as 15 to 20 transformer secondary windings have been connected in series to facilitate the operation of a long-distance transmission system, the maximum rating of each transformer being not greater than 10 kw. However, a method of constructing large transformers has long been devised by which an enormous amount of power may be transformed in a single unit. Such designs embody principles of insulation for high voltages and various methods for maintaining a low operating temperature. Ten years ago a 500-kilowatt unit was considered to be a large size of transformer. The history of ten years' development has shown a most interesting process of evolution: it has marked more than a tenfold increase in size up to the present day.

Going back to part of the history of transformer manufacturing we find that the first transformer used by Faraday in his historic experiments had for their magnetic circuit a closed ring of iron. Varley in the year 1856 pointed out the disadvantage of leaving the magnetic circuit open, gave it a closed path by bending back and overlapping the end of the straight iron wire core. In the early days of electric lighting Ferranti modified Varley's method by using, instead of iron wires, strips of sheet iron bent back and interlaced. The nearest approach to the present day practice was to embed link-shaped coils in the recesses of a core built up of shaped stampings, afterward completing the magnetic circuit either with sheets of laminations or with strips interlaced with the ends of the prejecting legs. There is good reason to believe that from this construction the "Shell" type transformer of to-day received its name.

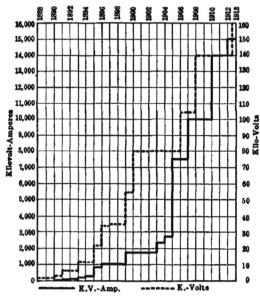

Fig. 1.—Transformer development.

Not very many years ago it was a question among very prominent engineers whether 15,000-volt transformers could be made to work.

Before the time engineers conceived the idea of drying transformer oil, it was not uncommon to see transformers without any solid material between coils—the oil space above being relied on for insulation. At that time the transformer was the only limiting feature of transmission. We have only to go back about 9 years to find the first 50,000- and 60,000-volt transformer actaully operating (see Fig. 1). At the present time we are actually operating transformers at 145,000 volts with as great safety and with less liability of break-down than formerly was experienced in operating at 15,000 volts, and the actual voltage limit of the transformer is not yet in sight, but on the contrary—the transmission line itself is the limiting feature in the voltage or the distance of transmission.

About 16 years ago the first 20,000-volt transformer was made. At the present time the commercial manufacture of 175,000-volt transformers is being considered. Fig. 1 represents the development of single capacity transformers up to the year 1913. Increased voltage means, of course, increased kv-a capacity of units, not only as regards the transformers themselves but in generators and prime movers.

Glancing back 25 years we enter upon a time when alternating currents were grievously fought in the Law Courts as being a current both *"dangerous and impracticable."*

Just about 20 years ago—the period when three-phase star and delta systems of electric distribution were recognized as practicable for commercial use—we find universal opposition to high voltages (above 1500 volts) and large units (above 500 kw.).

Going back only 10 years—about the period when 50,000 and 60,000 volts were recognized as practicable for transmission purposes—large commercial units and large commercial electric power systems, as recognized to-day, were universally considered as impossible, as absolutely unreliable or decidedly dangerous.

In view of the three above decades, practically accepted throughout the entire engineering world, we actually have at the present time electric power systems operating as *one company* and one centralized system delivering over 200,000 kw., and *single units* for commercial electric power purposes, as:

    Steam turbines (horizontal type) of 33,000 h.p.
    Steam turbines (vertical type) of 40,000 h.p.
    Water turbines (vertical type) of 20,000 h.p.
    Turbo-generators (horizontal type) of 25,000 kw.
    Turbo-generators (vertical type) of 30,000 k.w.
    Generators (vertical type) of 12,500 kw.
    Transformers (shell-type) of 14,000 kw.

and Transformers (three-phase group of three) of 18,000 kw.

Fundamental Principles.—The transformer consists primarily of three parts: the primary winding, the secondary winding, and the iron core. The primary winding is connected in one circuit, the secondary in another, and the core forms a magnetic circuit which links the two together.

STATIONARY TRANSFORMERS

The principle of the constant-potential transformer is easily explained if we neglect the slight effects of resistance drop in the windings, leakage of magnetic flux, and the losses. The primary winding is connected to a source of e.m.f., which connection would constitute a short-circuit were it not for the periodic changing in value which permits the flux produced by the current to generate a counter e.m.f. which holds the current down to a value just sufficient to produce that value of flux necessary to generate an e.m.f. in the primary and equal and opposite to the impressed e.m.f. This same flux is surrounded by the turns of the secondary winding, the same e.m.f. being generated in each turn of wire whether primary or secondary. If E_1 is the impressed e.m.f.

$$\frac{E_1}{N_1} = e = \text{e.m.f. per turn,}$$

wherein N_1 is the number of primary turns. Then if N_2 is the number of secondary turns,

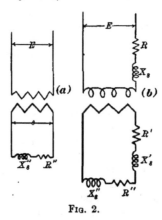

Fig. 2.

$$E_2 = N_2 e = \text{secondary e.m.f.}$$

and $$\frac{N_1}{N_2} = \frac{E_1}{E_2} = \text{ratio of transformation.}$$

When N_1 is greater than N_2, the transformer is called a "step-down transformer" and when N_1 is less than N_2 it is called a "step-up transformer."

The reader will understand that a step-up transformer may be

INTRODUCTION

used as a step-down transformer, or *vice versa*. The primary is the winding upon which the e.m.f. is impressed.

The primary and secondary windings of a transformer possess both resistance and reactance, and the secondary values may be reduced to primary terms by multiplying them by the square of the ratio of transformation. This applies to the load resistance and reactance as well. Thus, consider a circuit containing a transformer, a load, etc., as shown in Fig. 2. Obviously, to maintain the core flux a magnetizing current is required, which of course must pass through the primary winding.

(b) is an exact duplicate of the transformer (a). The factors R and X_s are respectively the resistance and reactance of the primary; $R'X'_s$ the respective resistance and reactance of the secondary winding, and R'' and X_s'' the load resistance and reactance. This representation is about the simplest for treatment of the transformer circuits.

Points in the Selection of Transformers.—The electrical characteristics of a transformer are mostly dependent upon the quality, arrangement and proportion of the iron and copper that enter into its construction. The losses are of two kinds: the copper loss, due to the current through the coils; and the iron loss, caused by the reversing of the magnetic flux in the core. These losses appear as heat, and suitable means must be provided for the disposal of this heat.

In selecting a transformer for a given service, it is advisable to consider first:

(a) The ratio of iron and copper loss, which should be such that the *total cost* of the losses is a minimum.

(b) The cost of the transformer for a given service and the cost of its total losses should be a minimum.

The cost of a transformer for a given service depends on the amount which must be paid for the losses during the life of the transformer and on the first cost of the transformer itself. In considering the losses and price paid for a transformer together, the losses may be conveniently represented as a capital cost by dividing their annual cost by the interest and depreciation factor.

Safety to life, durability, and economy are essential features of this apparatus in its ability to give continuous and uninterrupted service. These factors, sometimes in part and sometimes entire, are sacrificed to obtain a higher efficiency, especially in high voltage transformers where so much insulation has to be used. This is not considered good practice although the higher efficiency is obtained, and a transformer designed and built with the main object of efficiency at the expense of safety and reliability finally brings discredit to its maker. The loss in revenue alone due to the failure of a large power transformer would more than offset the saving of several years in gaining an additional fraction of 1 per cent. in efficiency, not counting the great loss of confidence and prestige on the part of the customer. The application of knowledge gained by many years of constant and careful study of all the properties and characteristics of transformers in actual practice has placed this type of apparatus on a plane which we may now call both safe and reliable for operaing voltages as high as 110,000 volts. Looking back over the development of the transformer we do not pass very many years before we enter the time when large units of moderately high voltage (20,000 volts) were considered by manufacturers a tedious if not altogether a dangerous undertaking, in fact quite as dangerous as designing and building a unit in these days to give an output of 20,000 kw. at 200,000 volts.

Cooling.—A well-designed transformer should not only maintain a low average temperature but the temperature should be uniform throughout all of its parts. The only effective way of insuring uniform temperature is to provide liberal oil ducts between the various parts of the transformer, and these should be so arranged in relation to the high and low voltage windings as to give the best results without sacrificing other important factors. Ducts necessarily use much available space and make a high voltage transformer of given efficiency more expensive than if the space could be completely filled with copper and iron; in view of the reliability and low deterioration of a transformer of this type, experience has demonstrated that the extra expense is warranted.

For various reasons the temperature rise in a transformer is limited. The capacity for work increases directly as the volume of material, and the radiating surface as the square of the dimensions; therefore, it is evident that the capacity for work increases faster than the radiating surface.

The amount of heat developed in a transformer depends upon its capacity and efficiency. For instance, in a 500-kw. transformer of 98 per cent. efficiency there is a loss at full load of about 7.5 kw.; and since this loss appears as heat, it must be disposed of in some way, or the temperature of the transformer will rise until it becomes dangerously high. This heat may be removed in several ways; by ample radiation from the surface of the tank or case in which the transformer is operated; by the circulation of water through pipes immersed in oil; or by the constant removal of the heated oil and its return after being cooled off.

The determination of the temperature may be made by thermometer or by the measurement of resistance. High temperature causes deterioration in the insulation as well as an increase in the core loss due to aging. The report of the Standardization Committee of the American Institute of Electrical Engineers specifies that the temperature of electrical apparatus must be referred to a standard room temperature of 25° C., and that a correction of 1/2 per cent. per degree must be made for any variation from that temperature, adding if less and subtracting if more.

The temperature rise may be determined by the change of resistance, using the temperature coefficient 0.39 per cent. per degree from and at zero degrees.

Whenever water is available and not expensive, water-cooled transformers are preferable to air-blast transformers of the large and moderate sizes (1,500 to 5,000 kw.), as it permits operation at lower temperatures and allows more margin for overloads. Where water is not available, there is a choice of two kinds of air-cooled transformers: the oil-filled self-cooled type, and the air-blast type which is cooled by a forced air circulation through the core and coils or by blowing air on the outer case of the transformer. This type of transformer is not very reliable for voltages above 35,000 volts, principally on account of the great thickness of the solid insulation needed and the consequent difficulty in radiating heat from the copper.

A great deal has been written about the fire risks of air-blast and oil-filled transformers, but this is a matter that depends as much on surrounding conditions and the location of the transformers as on their construction. The air-blast transformer contains a small amount of inflammable material as compared with the oil-cooled transformer, but this material is much more easily ignited. A break-down in an air-blast transformer is usually followed by an electric arc that sets fire to the insulation material, and the flame soon spreads under the action of the forced circulation of air; although the fire is of comparatively short duration it is quite capable of igniting the building unless everything near the transformer is of fireproof construction. The chance of an oil-filled transformer catching fire on account of a short-circuit in the windings is extremely small, because oil will only burn in the presence of oxygen, and, as the transformer is completely submerged in oil, no air can get to it. Moreover, the oil used in transformers is not easily ignited; it will not burn in open air unless its temperature is first raised to about 400° F., and with oil at ordinary temperatures, a mass of burning material can be extinguished as readily by immersing it in the oil as in water. In fact, the chief danger of fire is not that the oil may be ignited by any defect or arc within the transformer, but that a fire in the building may so heat the oil as to cause it to take fire. The idea of placing oil-filled transformers into separate compartments is not thought of so seriously as it was some years ago, although it does not apply in every case and therefore it is necessary to consider carefully when selecting this type of transformer.

Of the large number of factors in the make-up of a transformer only four which the operating manager is particularly anxious to know enter into the operating costs, namely: the core and copper losses, temperature, efficiency and regulation. All of these costs (since they might be called costs as they include the cost of generating such losses and of supplying the station capacity with which they generate them) represent quite a large amount of energy during the life of the transformer.

Losses.—The hysteresis and eddy current losses are generally combined under the term of "core loss," this loss occurring in all magnetic material which is subject to alternating magnetic stresses. The hysteresis loss, as is generally known, may be defined as the work done in reversing the magnetism in the steel, and it may be considered as due to the molecular friction from the reversal of magnetism, this friction manifesting itself as heat. The amount of hysteresis in a given steel varies with the composition, with the hardness, with the frequency of reversal of magnetism, with the maximum induction at which the steel is worked, and the temperature. The hysteresis loss varies approximately as the 1.6 power of induction, and directly as the frequency. The eddy current loss varies inversely as the ohmic resistance, directly as the square of the induction, and decreases as the temperature increases. It is greater in thick laminations than in thin (hysteresis being greater in hard steel than in soft

INTRODUCTION

steels), it is also greater as the insulation between adjacent laminations is less. Lowering the frequency of supply will result in increased hysteresis and higher temperature in the iron; reducing the frequency from say 133 to 125 cycles will entail an increased hysteresis of about 4 per cent., and a reduction from 60 to 50 cycles will raise the hysteresis approximately 10 per cent.

For equal output there will not exist any change in the copper loss, but in the case of large power transformers the increased temperatures due to excessive iron losses will materially decrease the output, and the normal rated secondary current or low voltage current will become a virtual overload.

Iron loss and exciting current in addition to decreased kw. capacity of the transformer mean greater coal consumption, all these factors being directly opposed to commercial operation, and as this iron loss is constant while the transformer is connected to the system, no matter what the load may be, the total yearly loss will represent a great loss in revenue. While the iron loss is practically constant at all loads, the copper or $I^2 R$ loss varies as the square of the current in both the high and low voltage windings. The output is the total useful energy delivered to the primary, and consists of the output energy plus the iron loss at the rated voltage and frequency plus the copper loss due to the load delivered. This loss is within easy control of the designer, as a greater or less cross-section of copper may be provided for the desired per cent. regulation. In a transformer core of a given volume and area, the number of turns for the required iron loss are fixed. To secure the desired copper loss advantage is taken of a form of coil wherein the mean length per turn is kept as low as possible with the necessary cross-section of copper. If the form of coil is rectangular it is evident that the mean length per turn of the conductor would be increased, provided the same cross-section or area of the core is enclosed, so in order to secure the shortest length per mean turn consistent with good construction it is necessary to adopt the square core in which the corners have been cut off. Also, in order that the greatest amount of conductor may be allowed for the available space, all wire entering into the low and high voltage windings is either square or rectangular in shape, as by using this form of conductor the area is increased by about 33 per cent. over that of ordinary round wire. This method permits the copper loss to be reduced, and at the same time allows a great part of the total copper loss to take place in the high voltage winding. The loss due to magnetic leakage is made negligible by virtue of the compact construction and the proper disposition of the windings with relation to each other and the core.

The copper loss generally has a less cost than the iron loss, due to the reduction in output charge because of its short duration, and also has a slightly less capital cost due to its diversity factor.

The losses due to the magnetizing current and heating are determined from manufacturers' guarantees or by test as the transformer is received from the factory. The exciting current of a transformer is made up of two components; one being the energy component in phase with the e.m.f. which represents the power necessary to supply the iron loss, the other component being in quadrature with the e.m.f. which is generally known as the magnetizing current and is "wattless" with the exception of a small $I^2 R$ loss. The magnetizing current has very little influence on the value of the total current in a transformer when it is operating at full load, but as the load decreases the effect of magnetizing current becomes more prominent until at no load it is most noticeable. The greater the exciting current, the greater is the total current at the peak of the load, and hence the greater must be the generating station equipment and transmission lines to take care of the peak.

Regulation.—It is often said that regulation reduces the voltage upon the load and therefore causes a direct loss in revenue by reducing the energy sold. If, however, the mean voltage with transformer regulation is maintained at the same value as the constant voltage without regulation, the energy delivered to the customer will be the same in both cases, hence there will be no direct loss of revenue. As the regulation of transformers is affected at high power factor mostly by resistance and at low power factor mostly by reactance, both should be kept as low as possible. With non-inductive load the regulation is nearly equal to the ohmic drop, the inductance having but little effect. With an inductive load the inductance comes into effect, and the effect of resistance is lessened; depending on the power factor of the load. In general, the core type transformer has not so good regulation as the shell type transformer; the reason for this is that in the shell type transformer there is a better opportunity for interlacing the coils.

Core Material.—Several grades of steel are manufactured for transformers. Certain peculiar ingredients are added to the pure iron in such proportions and in such a manner that the resultant metal is, actually speaking, neither iron nor steel, and for the want of a better name it is termed by the trade an "alloy steel." The effect of various substances, such as silicon, phosphorus, sulphur, etc., has long been a matter of common knowledge among those familiar with the metallurgy of steel, but the electro-magnetic properties of some of the late "alloy steels" have been known a comparatively short time. Some steels are very springy and resist bending, but ordinary steels are comparatively soft and yielding, and easily crease when bent. During the past few years a great deal of time has been spent in experimental work to determine the best shape of core and the proportioning of the various elements of a transformer to give the highest efficiency with minimum cost.

If a transformer had a perfect magnetic iron circuit no losses would occur due to imperfect iron, etc. Losses do occur and increase with the aging of the iron. The cost of this iron loss will include the cost of generating such loss and of supplying the station capacity with which to generate it, and it also includes the cost of transmitting the energy consumed by the loss from the generating station to the transformer. The revenue affected by imperfect iron of a low grade put into the transformers operating at the end of a long-distance transmission line may be divided up as follows:

Cost of iron losses in the transformers.

Cost of this energy passing over the transmission line, affecting both efficiency and regulation for the same amount of copper.

Extra cost at the generating station in generators, transformers, etc., to take care of this energy.

Cost of magnetizing current in the transformers.

Cost of this *additional* current passing over the transmission line.

Extra cost at the generating station in generators, transformers, etc., to take care of this extra current.

Insulation.—While the quality, arrangement and proportion of the iron and copper are essentials in transformer design, the proper selection, treatment and arrangement of the insulating material require even greater skill and wider knowledge than does the proportioning of copper and iron. A transformer will not operate without sufficient insulation, and the less space occupied by this insulation the more efficient will it be, with a given amount of iron and copper.

At the present time the use of solid compounds for impregnation of the winding of transformers is almost universally adopted. The use of this compound marks a great improvement in the modern transformer because it helps to make the coils mechanically stronger by cementing together the turns and the insulation between turns and layers in the windings.

All high-voltage transformers and practically all transformers of any voltage, are dried and impregnated by the vacuum process which is now considered to be the most reliable insulating material and method of insulating. For this purpose both asphalts and resins are the materials available. They can be liquified by heat and forced into the coil in that condition, and on cooling they harden, forming a solid mass (coil and material) which is, if well done, free from porosity and volatile solvents. The compound fills the porous covering of the wire-conductors, and all other spaces in the coils no matter how small, thus increasing the dielectric strength and preventing moisture from soaking into the coils. Before applying this process, the coils are thoroughly dried either in a separate oven or in the impregnating tank. They are then placed in the impregnating tank and heat is applied until the coils reach a temperature at which the impregnating compound is thoroughly fluid. The air in the tank is then exhausted by a vacuum pump. After the vacuum has exhausted the last traces of moisture from the coils, hot compound from another tank is drawn into the impregnating tank until the coils are thoroughly covered, this condition being maintained until the coils are impregnated. The pressure generally used is 60 to 80 lb. per square inch. The time required under vacuum and pressure can only be determined by trial but usually from three to six hours' vacuum will dry any ordinary high-voltage coil not unduly moist.

At the present time the fluid point of some impregnating compounds is about 95° C., but it is possible that the development of synthetic gums will soon reach a stage which will permit an actual operating temperature of 130° C.

The National Board of Fire Underwriters specify that the insulation of nominal 2000-volt transformers when heated shall withstand continuously for one minute a difference of potential

INTRODUCTION

of 10,000 volts (alternating) between primary and secondary coils and the core and a no-load run of double voltage for 30 minutes.

All transformers should be subjected to insulation tests between the primary and secondary, and the secondary and core. A transformer may have sufficient strength to resist the strain to which it is constantly subjected, and yet due to an imperfection in the insulation may break down when subjected to a slight over-voltage such as may be caused by the opening of a high-power circuit. The application of a high-potential test to the insulation will break down an inferior insulation, or a weak spot or part of the structure in the insulation. The duration of the test may vary somewhat with the magnitude of the voltage applied to the transformer. If the test be a severe one, it should not be long continued, for while the insulation may readily withstand the application of a voltage five or even six times the normal strain, yet continued application of the voltage may injure the insulation and permanently reduce its strength.

FUNDAMENTAL EQUATIONS

In the design of successful transformers, the following equations are found reliable:

Let N = Total number of turns of wire in series.
ϕ = Total magnetic flux.
A = Section of magnetic circuit in square inches.
f = Frequency in cycles in seconds.
B = Lines of force per square inch.
E = Mean effective e.m.f.

$$4.44 = \frac{2\pi}{\sqrt{2}} = \sqrt{2} \times \pi$$

then
$$E = \frac{4.44 \, f \, \phi \, N}{10^8} \qquad (1)$$

Equation (1) is based on the assumption of a sine wave of e.m.f., and is much used in the design of transformers.

If the volts, frequency, and number of turns are known, then we have

$$\phi = \frac{E \times 10^8}{4.44 \times f \times N} \qquad (2)$$

14 STATIONARY TRANSFORMERS

If the volts, frequency, cross-section of core, and density are known, we have:

$$N = \frac{E \times 10^8}{4.44 \times f \times B'' \times A} \quad (3)$$

Magnetic densities of transformers vary considerably with the different frequencies and different designs.

Current densities employed in transformers vary from 1000 to 2000 circular mils per ampere.

Efficiency.—The efficiency of a transformer at any load is expressed as:

$$\text{Efficiency} = \frac{\text{output}}{\text{output} + \text{core loss} + \text{copper loss}} \quad (4)$$

In the case of ordinary transformers with no appreciable magnetic leakage, the core loss is practically the same from no-load to full load. The only tests required, therefore, in order to obtain the efficiencies of such transformers at all loads, with great accuracy, are a single measurement by wattmeter of the watts lost in the core, with the secondary on open circuit; and measurements of the primary and secondary winding resistances, from which the $I^2 R$ watts are calculated for each particular load. The core loss, which is made up of the hysteresis loss and eddy-current loss, remains practically constant in a constant-potential transformer at all loads. In the case of constant-current transformers and others having considerable magnetic leakage when loaded, this leakage often causes considerable loss in eddy currents in the iron, in the copper, and in the casing or other surrounding metallic objects. It should be borne in mind that the efficiency will also depend on the frequency and the wave-form and that the iron core may age; that is to say, the hysteresis coefficient may increase after the transformer has been in use some time. Generally speaking, the efficiency of a transformer depends upon the losses which occur therein, and is understood to be the ratio of its net output to its gross power input, the output being measured with non-inductive load.

All-day Efficiency.—The point of most importance in a transformer is economy in operation, which depends not only upon the total losses, but more particularly upon the iron or core loss. For example, taking two transformers with identical total losses, the one showing the lower iron loss is to be preferred, because of the greater all-day efficiency obtained, and

INTRODUCTION

the resulting increase in economy in operation. This loss represents the energy consumed in applying to the iron the necessary alternating magnetic flux, and is a function of the quality of the iron and the flux density at which it is worked, or in other words, the number of magnetic lines of force flowing through it.

The all-day efficiency mentioned above is the ratio of the total energy used by the customer, to the total energy input of the transformer during twenty-four hours. The usual conditions of present practice will be met, if based on five hours at full load and nineteen hours at no load; therefore, "all-day" efficiency can be obtained from the following equation:

$$\text{All-day efficiency} = \frac{\text{Full load} \times 5}{\text{Core loss} \times 24 + I\,R \times 5 + \text{Full load} \times 5} \quad (5)$$

The importance to the economical operation of a central station of testing for core loss every transformer received, cannot be overestimated. The variation in core loss of two transformers of identical design may be considered, as depending, not only upon the constituents of the steel used, but also upon the method of treatment.

It has been found in practice, that transformers having initially low iron losses, after being placed in service, would show most decided increase under normal conditions. This increase is due to the "aging" of the iron. The aging of iron depends on the kind of material used, and on the annealing treatment to which it has been subjected. It has been shown that when steel is annealed so as to have a low loss, and then subjected to a temperature of from 85°C. to 100° C., the loss usually increases, in some cases this increase being as much as 300 to 400 per cent.

By some manufacturers of transformers it is claimed that the steel used in their cores is non-aging, or that it has been artificially aged by some process. However, it should be remembered that an absolutely non-aging steel is not as yet a commercial possibility. Within very short periods the iron losses sometimes increase, and under very high temperature conditions the laminations will become tempered or hardened, whereby the permeability is greatly reduced; therefore, the iron losses increase with the length of time the transformer is in operation.

The other factor affecting the efficiency of a transformer is the copper loss. It occurs only when the transformer is loaded, and while it may be considerable at full load it decreases very rapidly as the load falls off. As the transformer is seldom operated at full load, and in many cases supplies only a partial load for a few hours each day, the actual watt-hours of copper loss is far below the actual watt-hours of iron loss. However, for equal full-load efficiencies, the transformer having equal copper and iron losses is cheaper to manufacture than one in which the iron loss is reduced, even though the copper loss is correspondingly increased.

Regulation.—The ability of a distribution transformer to deliver current at a practically constant voltage regardless of the load upon it, is a very highly important feature. By the use of conductors of large cross-section and by the proper interlacing of primary and secondary coils, extremely close regulation may be obtained with loads of various power-factors, thus tending to lengthen the life of lamps and to improve the quality of the light.

In well-designed transformers, low core loss and good regulation are in direct opposition to one another yet both are desired in the highest degree. The regulation of a transformer is understood to be the ratio of the rise of secondary terminal voltage from full load to no-load, at constant primary impressed terminal voltage, to the secondary terminal voltage. In addition to the vastly improved service, it is possible to adopt the efficient low-consumption lamp, when the transformers in use maintain their secondary voltage at a practically constant value when the load goes on or off. While so few central stations are able to keep their voltage constant within 2 per cent. it may be concluded that at present the point of best practical regulation on transformers from about 5 kw. up, lies between the values of 1.75 per cent. to 2.00 per cent.

Regulation is a function of the ohmic drop and the magnetic leakage. To keep the iron loss within necessary limits and at the same time secure good regulation is an interesting problem. We may reduce the resistance of the windings by using fewer turns of wire, but with fewer turns the iron is compelled to work at a higher flux density, and consequently with an increased loss. If we adopt a larger cross-section to reduce the flux density we need a greater length of wire for a given number of turns which

INTRODUCTION

thus gives an increase in resistance. The remaining expedient only is to use a larger cross-section of copper, while keeping down the flux density by employing a sufficient number of turns, to secure the low resistance necessary for good regulation. For ordinary practice the regulation of a transformer for non-inductive loads may be calculated as follows:

$$\% \text{ regulation} = \% \text{ copper loss} - \frac{(\% \text{ reactance drop})^2}{200} \quad (6)$$

For inductive loads the regulation may be calculated by the following equation:
Per cent. regulation =

$$\frac{\text{per cent. reactance drop}}{\sin \theta} + \frac{\text{per cent. resistance drop}}{\cos \theta} \quad (7)$$

wherein θ is the angle of phase displacement between the current and the e. m. f.

Regulation on inductive loads is becoming more important as the number of systems operating with a mixed load (lamps and motors) is constantly increasing. Many transformers while giving fair regulation on non-inductive loads, give extremely poor regulation on inductive loads.

VECTOR REPRESENTATION

In studying the performance of transformers it is simple and convenient to use graphical methods. The graphical method of representing quantities varying in accordance with the sine law has been found to be one of the simplest for making clear the vector relations of the various waves to one another.

The principle of this method is shown in Fig. 3, where the length of the line $o\ e$ represents the magnitude of the quantity involved, and the angle $e\ o\ x = \theta$ represents its phase position either in time or space.

In an alternating-current circuit the relation between the most important quantities may be represented by the method above mentioned. When such diagrams are used to represent voltages or currents, the length of the lines represents the scale values of the quantities, while the angles between the lines represent the angle of phase difference between the various quantities. The diagrams are constructed from data available in each case. The diagram below represents a circuit containing resistance and inductive reactance. Since the $I\,R$ drop is always in phase with the current and the counter e.m.f. of

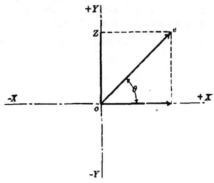

FIG. 3.—Vector diagram.

self-inductance in time-quadrature with the current which produces the m.m.f., these two magnitudes will be represented by two lines, $o\,e_1$ and $o\,e_3$, at right angles to each other; their

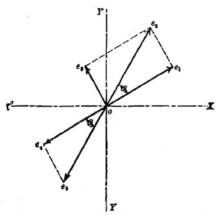

FIG. 4.—Vector diagram of a transformer, assuming an inductive load.

sum being represented by $o\,e_2$, representing the resultant value of these two e.m.fs., and is, therefore, equal and opposite to the e.m.f., which must be impressed on the circuit to produce the

INTRODUCTION

current, I, against the counter e.m.f. of self-inductance, $2\pi f L\ I$, and the counter e.m.f. of resistance, $I\ R$. Therefore, by the properties of the angles,

$$e_s{}^2 = (I\ R)_2 + (2\pi f L\ I)^2 \tag{8}$$

The angle of lag of the current behind the e.m.f. is shown as the angle between the lines representing the resistance e.m.f. and that representing the resultant of the resistance and the reactance e.m.f. Since the resistance component of the impressed e.m.f. is in phase with the current and differs 180 degrees in phase from the resistance e.m.f., its position will be that shown by the line $o\ e_4$, and the angle between that line and $o\ e_5$ is the angle of lag.

The tangent of the angle, θ, is equal to $\dfrac{o\ e_2}{o\ e_1}$.

$o\ e_3 = 2\pi f L\ I$ and $o\ e_1 = I\ R$,

$$\text{or } \tan \theta = \frac{2\pi f L\ I}{I\ R} = \frac{2\pi f L}{R} \tag{9}$$

Denoting $I = I$.

$$(2\pi f L = X_s = \omega L. \quad = j\,x_s) \tag{10}$$

Like the e.m.f., the current values can be split into two components, one in phase and one in quadrature with the e.m.f., the same results being obtained in both cases.

Impedance is the apparent resistance in ohms of the transformer circuit, and is that quantity which, when multiplied with the total current will give the impressed volts, or $I\ Z = E$

Denoting Z as $R + j\ x$; where j is an imaginary quantity

$$\sqrt{-I.}, \text{ or } Z = R + \sqrt{-1}\ Lw \tag{11}$$

In measuring the energy in an alternating current circuit it is not sufficient to multiply E by I as in the case of direct current, because a varying rate of phase between the voltage and current has to be taken into account. This phase angle can be determined by a voltmeter, an ammeter and a wattmeter, and is expressed as

$$\text{Cos. } \phi = \frac{P}{E\ I} \tag{12}$$

where P is the actual power in watts consumed by the load;

EI the apparent watts, or I^2R as it is sometimes called, and $\cos\phi$ the angle of phase displacement.

It is very evident that when the resistance is large compared

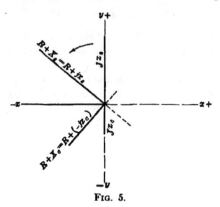

Fig. 5.

with the reactance, the angle of time-lag is practically zero. (See Fig. 174.) If the reactance is very large compared with the resistance, the angle of lag will be almost 90 time-degrees; in other words, the current is in quadrature with the e.m.f.

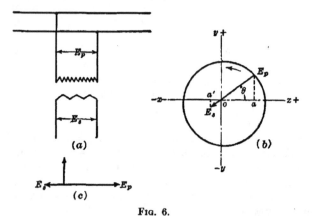

Fig. 6.

A problem which can always be solved by the use of transformers, is the convertion of one polyphase system into another. Since in the original system there must be at least two components of e.m.f. which are displaced in time-phase, by varying the values of these components a resultant of any desired phase can be obtained. In phase-splitting devices using inductive or condensive reactance, an e.m.f. in quadrature with the impressed e.m.f. is obtained from the reactive drop of the current through an inductive winding, or a condenser, and the necessary energy is stored as magnetic energy, $\frac{I^2 L}{2}$, in the core of the winding, or as electrostatic energy, $\frac{E^2 C}{2}$, in the dielectric of the condenser; but such devices are of little practical use.

CHAPTER II

SIMPLE TRANSFORMER MANIPULATIONS

There are a number of different ways of applying transformers to power and general distribution work, some of which are:

Single-phase (one, two, three or more wire).
Two-phase (three, four, five or more wire).
Three-phase delta (grounded or ungrounded).
Three-phase star (grounded or ungrounded).
Three-phase Tee (grounded or ungrounded).
Three-phase open-delta.
Three-phase star to star-delta or *vice versa*.
Three-phase star and delta or *vice versa*.
Three-phase to two-phase or *vice versa*.
Three-phase to two-phase-three-phase or *vice versa*.
Three-phase to six-phase, or *vice versa*.
Two-phase to six-phase, or *vice versa*.
Three-phase to single-phase.
Two-phase to single-phase.

The principal two precautions which must be observed in connecting two transformers, are that the terminals have the same polarity at a given instant, and the transformers have practically identical characteristics. As regards the latter condition, suppose a transformer with a 2 per cent. regulation is connected in parallel with one which has 3 per cent. regulation; at no load the transformers will give exactly the same e.m.f. at the terminals of the secondary, but at full load one will have a secondary e.m.f. of, say, 100 volts, while the other has an e.m.f. of 99 volts. The result is that the transformer giving only 99 volts will be subject to a back e.m.f. of one volt, which in turn will disturb the phase relations and lower the power-factor, efficiency and combined capacity; in which case it is much better to operate the secondaries of the two transformers separately. In order to determine the polarity of two transformers proceed with the parallel connection as if everything were all right, but connect the terminals together through two small strips of fuse wire, then close the primary switch. If the fuse blows, the connections must be reversed; if it does not, then the connections may be made permanent.

The primary and secondary windings of transformers may be connected to meet practically any requirement. Fig. 6 represents the ordinary method of connecting a single-phase transformer to a single-phase circuit. Referring to the graphical representation in Fig. 6 it is shown that $O\,E\,p$ and $O\,E\,s$ (the primary and secondary e.m.fs.) represent two lines of constant length, rotating at a uniform rate about O as a center. The direction of the secondary is not strictly 180 electrical degrees out of time phase with the primary, but for convenience and elementary purposes it is commonly represented as such. The dotted line is vertical to $O\,X$, so that as the points $E\,p$ or $E\,s$ move

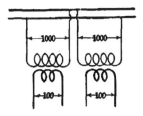

Fig. 7.—Straight connection of two ordinary single-phase transformers.

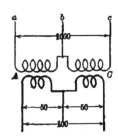

Fig. 8.—Single-phase transformer with primary and secondary coils both in series.

in the circle, they occupy variable distances from O. As they travel from X through Y, it is evident that they have positive and negative values, and that these values vary from zero to a definite maximum. They pass through a complete cycle of changes from positive to negative and back to positive, corresponding to a complete revolution, both the e.m.fs varying as the sine of the angle $O = O\,E\,p\,A$.

Since the changes of voltages in the primary and secondary windings of a transformer go through their maximum and minimum values at the same time, the result of connecting the two windings in series is to produce a voltage which is either the sum or the difference of the voltages of the windings, according to the mode of joining them. If the windings of a step-up transformer are joined in series so that their resultant voltage is the sum of the voltages of the two windings, the source of supply may be connected to the terminals of the composite winding, instead of at the terminals of what was originally the primary. If this is done, the windings of the transformer may be reduced until the total voltage of the two windings equals the voltage of the original primary winding.

Fig. 7 shows the way in which two ordinary single-phase transformers are connected.

Fig. 8 shows one transformer which has two secondary coils connected in series. If this transformer be of the core type and the two coils arranged on different limbs of the core, it will be advisable to have the fuse in the middle wire considerably smaller than the fuses on the two outside wires. The reason for this is, that should one of the fuses on the outside circuits blow, say, for instance the fuse on leg A, the secondary circuit through this half-section will be open-circuited, and the primary coil corresponding to this section will have a greater impedance than the other half of the coil, the inductance of which will be neutralized by the load on the other half of the secondary coil. The result will be that the counter e.m.f. of the primary section, A, will be greater than that of section C, because the two sections are in series with each other, and the current must be the same in both coils; therefore, the difference of potential between the primary terminals, A, will be greater than that between the primary terminals of C, consequently the secondary voltage of C will be greatly lowered.

Manufacturers avoid the above mentioned disadvantage by dividing each secondary coil into two sections, and connecting a section of one leg in series with a section of the coil on the other leg of the core, so that the current in either pair of the secondary windings will be the same in coils about both legs of the core.

Transformers are made for three-wire service having the windings so distributed that the voltage on the two sides will not differ more than the regulation drop of the transformer, even with one-half the rated capacity of the transformers all on one side; with ordinary distribution of load the voltage will be practically equal on the two sides.

Fig. 9 shows a single-phase transformer with two coils on the primary, and two coils on the secondary. The primaries are shown connected in parallel across the 1000-volt mains, and the secondaries are also connected in parallel.

To obtain a higher secondary voltage the coils may be con-

SIMPLE TRANSFORMER MANIPULATIONS 25

nected as shown in Fig. 10. In this case the primary coils are connected in parallel, and the secondary coils connected in series. The difference of potential across the two leads, with the primaries connected in parallel and the secondaries connected in series will be 200 volts, or 100 volts per coil.

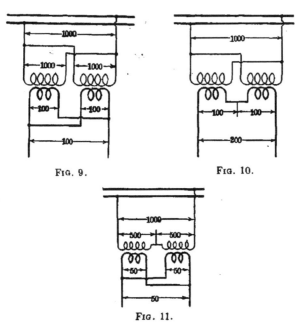

Fig. 9.
Fig. 10.
Fig. 11.

Fig. 9.—Transformer with primary and secondary windings both in parallel.

Fig. 10.—Transformer with primary windings connected in parallel and secondary windings in series.

Fig. 11.—Transformer with primary windings connected in series and secondary windings in parallel.

Note—For convenience all ratios of transformation will be understood to represent ten to one (10 to 1).

If we invert the arrangement shown in Fig. 10 by connecting the primary coils in series, and connecting the secondary coils in parallel, we shall obtain a secondary voltage of 50, as represented in Fig. 11.

In Figs 12- 12ᵃ and 13 are represented a right and wrong way of connecting transformers in series or parallel, just as the case may be. The connections shown in Fig. 12 represent the right way of connecting two transformers in parallel, or in series, the solid lines showing the series connection and the dotted lines the parallel connection.

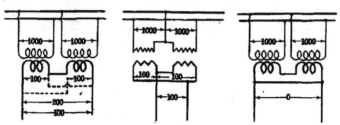

Fig. 12.—The right way of connecting single-phase transformers in parallel.

Fig. 13.—The wrong way of connecting single-phase transformers in parallel.

The connections shown in Fig. 13 are liable to happen when the transformers are first received from the factory. Through carelessness, the leads are often brought out in such a manner as to short-circuit the two coils if connected as shown. In this case the sudden rush of current in the primary windings would burn out the transformer if not protected by a fuse.

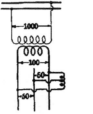   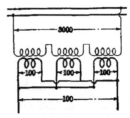

Fig. 14.—Three-wire secondary distribution.

Fig. 15.—Three 1000-volt transformers connected in series to a 3000-volt circuit.

The three-wire arrangement shown in Fig. 14 differs in every respect from the three-wire system represented in Figs. 8 and 10. The two outside wires receive current from the single-phase transformer, and the center, or neutral, wire is taken care of by a balancing transformer connected up at or near the center of distribution. The balancing transformer need only be of very small size, as it is needed merely to take care of the variation of load between the two outside wires.

It is sometimes desirable to use a much higher voltage than that for which the transformers at hand have been designed, and to attain this, the secondary wires of two or more transformers may be connected in parallel, while the primary wires may be connected in series with the source of supply.

This manner of connecting transformers is shown in Fig. 15. It, however, involves a high-voltage strain inside the separate transformers, between the high- and low-tension windings, and is therefore used only in special cases of necessity.

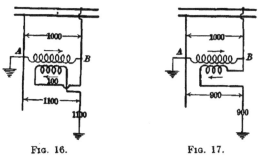

Fig. 16. Fig. 17.

Fig. 16.—Connection between primary and secondary windings, which gives 1100 volts across the secondary distribution wires. Boosting transformer.

Fig. 17.—Connection between primary and secondary windings from which we obtain 900 volts. Lowering transformer.

While it is possible to insulate for very high voltages, the difficulties of insulation increase very rapidly as the voltage is raised, increasing approximately as the square of the voltage.

Consider the case of a single-phase transformer as shown in Fig. 7. There is evidently a maximum strain of 1000 volts from one high-tension line wire to the other, and a strain of 500 volts from one line wire to ground, if the circuits are thoroughly insulated and symmetrical. The strain between high-tension and low-tension windings is equal to the high-tension voltage, plus or minus the low-tension voltage, depending on the arrangement and connection of the coils. With the arrangement shown in Fig. 16, it is quite possible to obtain 1100 volts between

. the wire, B, and ground, and the first indication of any such trouble is likely to be established by a fire, or some person coming in contact with a lamp socket, or other part of the secondary circuit that is not sufficiently insulated.

Should a ground exist, or in other words, a short-circuit between the high-tension and low-tension windings, it will, in general, blow fuses, thus cutting the transformer out of service; or the voltage will be lowered to such an extent as to call attention to the trouble, but the secondary windings must be grounded.

To avoid this danger to life, the grounding of secondaries of distribution transformers is now advocated by all responsible

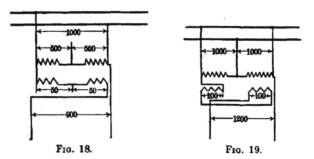

Fig. 18.   Fig. 19.

electric light and power companies in America. Differences of opinion have arisen as to general details both as regards the scope of grounding and the methods to be employed, but there is nevertheless a decided and uniform expression that the low voltage secondaries of distribution transformers should be permanently and effectively grounded.

The National Electrical Code provides for the grounding of alternating-current secondaries for voltages up to 250 volts. This voltage was decided upon after extensive investigation and discussion.

The principal argument for this grounding is the protection of life. A fault may develop in the transformer itself, between the primary and secondary wires which are usually strung one set above the other (high voltage always above the low voltage conductors), or a foreign circuit conductor such as an electrified series arc or incandescent lighting line conductor may come into metallic contact with a secondary line wire leading from transformer. Both the secondary line conductors and secondary transformer windings are liable to several forms of faults and danger due to a high voltage.

Some of the accidents recorded and due to such causes, are the handling of portable incandescent lamps and switching on lights

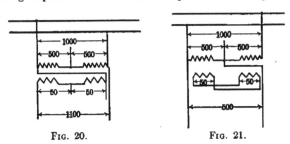

Fig. 20.   Fig. 21.

located in rooms with tile, cement or stone floors, by means of the switches attached to lamp sockets.

Fig. 17 represents an arrangement of primary and secondary circuits that may accidentally be made. These conditions immediately establish a potential difference of 900 volts. A great number of other single-phase transformer combinations may be used, some of which are shown in Figs. 18, 19, 20, 21, and 22.

If we take a transformer with a ratio of 10 to 1, say 500 and 50 turns respectively, and join the two windings in series, we find the number of turns required is only a total of 500 if the voltage is applied to the ends of the whole winding since the ratio of primary to secondary turns still remains 10 to 1, and with the same magnetic induction in the core, the primary counter e.m.f. and the secondary e.m.f. will both remain exactly as before; the ratio of primary and secondary currents remains also the same as before but is not produced in quite the same manner since the primary current will flow into the secondary and take the place of part of the current which would have been induced in the first case where the windings are separated.

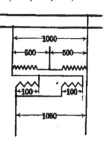

Fig. 22.

CHAPTER III

TWO-PHASE TRANSFORMER CONNECTIONS

So far as transformers are concerned in two-phase distribution, each circuit may be treated independently of the other as shown in Fig. 23, which is connected as though each primary and secondary phase were only a straight, single-phase system. One transformer is connected to one primary phase to supply one secondary phase, independent of the other phase, and the other transformer is connected to the other primary phase, supplying the other secondary phase.

In the two-phase system the two e.m.fs. and currents are 90 time-degrees or one-fourth of a cycle apart. The results which may be obtained from various connections of the windings of single-phase transformers, definitely related to one another in point of time, may be readily determined by diagrams.

The vector diagram in which e.m.fs. and currents are represented in magnitude and phase by the length and direction of straight lines, is a common method for dealing with alternating-current phenomena. To secure a definite physical conception of such diagrams, it is useful to consider the lines representing the various e.m.fs. and currents, as also representing the windings which are drawn, to have angular positions corresponding to angles between the lines; the windings are also considered to have turns proportional in number to the length of the corresponding lines and to be connected in the order in which the lines in the diagrams are connected.

The method of connecting two transformers to a four-wire, two-phase system is shown in Fig. 24. Both phases, as will be seen, are independent in that they are transformed in separate transformers.

A method of connection commonly used to obtain economy in copper is that shown in Fig. 24 where the primaries of the transformers are connected independently to the two phases, and the secondaries, are changed into a three-wire system, the center, or neutral wire being about one-half larger than each of the two outside wires.

TWO-PHASE TRANSFORMER CONNECTIONS

When two transformers having the same ratio are connected in parallel with a common load, the total secondary current is divided between them very nearly in inverse proportion to their impedances. This inverse impedance is usually expressed as

$$Y = \frac{P}{Z} = \frac{P}{R + \sqrt{-1}\, Lw} \tag{13}$$

Consider, for instance, a 5 kv-a transformer with an impedance of 2.9 per cent. and a 4 kv-a transformer having an impedance of 2.3 per cent. The admittances will be

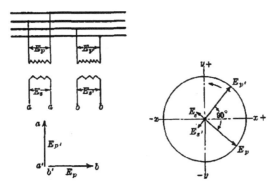

Fig. 23.—Two-phase four-wire arrangement.

$$\frac{5}{2.9} = 1.72 \text{ ohms and } \frac{4}{2.3} = 1.74 \text{ ohms}$$

The division of a total load of 9 kv-a on these when connected directly in parallel is

for the 5 kv-a $\frac{1.72 \times 9}{3.46} = 4.57$ kv-a or 91.5 per cent. rated load.

and

for the 4 kv-a $\frac{1.74 \times 9}{3.46} = 4.53$ kv-a or 113 per cent. rated load.

the value 3.46 is the total sum of admittances or $1.74 + 1.72 = 3.46$ ohms.

Lighting transformers are generally not mounted closely together when parallel operation is required, but are usually on a secondary net-work. In such cases where the drop due to the resistance of wiring or load between the two transformers is considerable, any difference such as ordinarily exists between different designs and different sizes would usually be automatically compensated for, so that the transformers would each take their proper proportion of load.

Assume these same two transformers are connected in parallel at a distance of about 500 ft. apart. Assume also that the center of load is 200 ft. from the 5 kv-a transformer and that the secondary wiring consists of a No. 0 wire. Neglecting altogether the reactance which will be small, as wires will doubtless be fairly close together, the drop due to resistance from the 4 kv-a transformer to the center of load will be 1.94 per cent., and from the 5 kv-a transformer about 1.62 per cent. Adding these resistances to the resistance component of the impedance of the two transformers, the impedance of the 5 kv-a transformer will be increased to 4.15 per cent. and the 4 kv-a transformer 4.01 per cent. The division of the total load will be

for the 5 kv-a $\frac{1.2 \times 9}{2.2} = 4.92$ kv-a or 98.5 per cent. rated load.
and
for the 4 kva $\frac{1.0 \times 9}{2.2} = 4.08$ kv-a or 102 per cent. rated load.

the value 2.2 being the sum of the admittances $1.2 + 1.0 = 2.2$. ohms.

So long as the two transformers are not connected in parallel it makes no difference which secondary wire of any one of the two transformers is connected to a given secondary wire. For example: It is just as well to connect the two outside wires, a and b, together, as it is to connect a' and b' as shown in Fig. 24. However, it makes no difference which two secondary wires are joined together, so long as the other wires of each transformer are connected to the outside wires of the secondary system. The two circuits being 90 time-degrees apart, the voltage between a and b is $\sqrt{2} = 1.141$ times that between any one of the outside wires and the neutral, or common return wire. The current in c is $\sqrt{2} = 1.141$ times that in any one of the outside wires.

Fig. 25 shows another method of connecting transformers, where the common return is used on both primary and secondary. With this method, there is an unbalancing of both sides of the system on an induction-motor load, even if all the motors on the system should be of two-phase design. The unbalancing is due to the e.m.f. of self-induction in one side of the system being in phase with the effective e.m.f. in the other side, thus affecting the current in both circuits.

Various combinations of the two methods shown in Figs. 16 and 18 can be made by connecting the primaries and secondaries

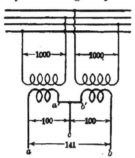

FIG. 24.—Two-phase four-wire primary with three-wire secondary.

as auto-transformers similar to the single-phase connections just mentioned.

The four ordinary methods of connecting transformers are represented by vectors in Fig. 26, a, b, c and d, showing relative values of e.m.fs. and currents.

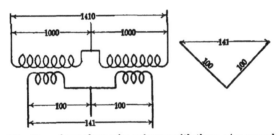

FIG. 25.—Two-phase three-wire primary with three-wire secondary.

$E_{ab} = 2e$	$E_{ab} = 2e$
$E_{a'b'} = 2e$	$E_{aa'} = \sqrt{2}e$
$I = i$	$I = i$
(A)	(B)
$E_{ab} = e$	$E_{ab} = 2e$
$E_{a'a} = \sqrt{2}e$	$E_{aa'} = 2\sqrt{2}e$
	$E_{a'b} = 2e$
$I_a = \sqrt{2}i$	I_a and $b = i$
	$I_{bb'} = \sqrt{2}i$
(C)	(D)

Let E = impressed volts per phase for A, B and D.

$e = \dfrac{E}{2}$ impressed voltage per phase for A, B and D.

$e = E$ impressed voltage per phase for C.

$I = i$ = current per phase for A and B.

Ic = current times $\sqrt{2}$ per phase for D.

Ia', Ib, Ia, $Ib' = \sqrt{2}$ times the current per phase for C.

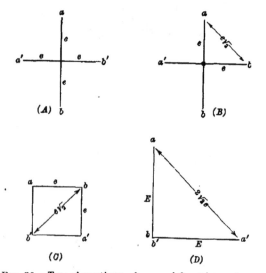

Fig. 26.—Two-phase three-, four- and five-wire systems.

Another arrangement is to connect the middles of the two transformer secondaries, as shown in Fig. 27. This method gives two main circuits, ac and df, and four side circuits, ad, dc, cf, and fa. The voltage of the two main circuits, between d and f is 100, and between a and c is 100. But the voltage across any one of the side circuits is one-half times that in any one of the main circuits, times the square root of two, or $50 \times \sqrt{2} = 70$ volts.

Another method shown in Fig. 28, commonly called the five-wire system, is accomplished by connecting the secondaries at the middle, similar to the arrangement in Fig. 27, and bringing out an extra wire from the center of each transformer.

The difference of potential between a and e will be $100 \times \sqrt{2} =$

141 volts, that across $b\ d$ will be $50\times\sqrt{2}=70$ volts, and that across any one of the main circuits will be 100 volts.

Another very interesting two-phase transformation may be obtained from two single-phase transformers by simply connecting the two secondary windings together at points a little to one side of the center of each transformer (see Fig. 29). There

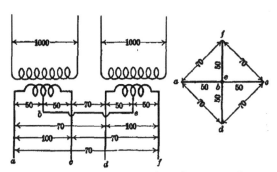

Fig. 27.—Two-phase star or four-phase connection.

are to be obtained $75\times\sqrt{2}=106$ volts between a and f; $\sqrt{75^2+25^2}=79$ volts between a and d, and c and f; $25\times\sqrt{2}=35$ volts between d and c, and 100 volts across each of the secondary windings.

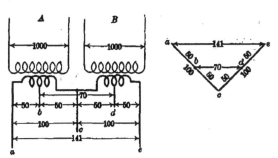

Fig. 28.—Two-phase five-wire secondary distribution.

It is possible by a combination of two single-phase transformer connections, to change any polyphase system into any other polyphase system, or to a single-phase system.

The transformation from a two-phase to a single-phase system is effected by proportioning the windings; or one transformer may be wound for a ratio of transformation of 1000 to 50; the other a ratio of 1000 to 86.6, or ($\sqrt{\frac{3}{2}}$). The secondary of this transformer is connected to the middle of the secondary winding of the first.

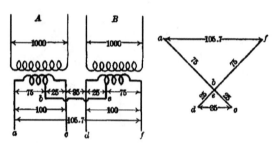

Fig. 29.—Two-phase multi-wire distribution.

In Fig. 30, $a\,c$ represents the secondary potential from a to c in one transformer. At the angle of 90 degrees to $a\,c$ the line, $c\,d$, represents in direction and magnitude the voltage between c and b of the other transformer. Across the terminals, $a\,c$, $c\,b$, and $a\,b$, it follows that three e.m.fs. will exist, each differing in

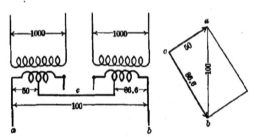

Fig. 30.—Two-phase to single-phase distribution.

direction and value. The e.m.f. across $a\,b$ is the resultant of that in $a\,c$ and $c\,b$ or 100 volts.

A complete list of two-phase, three- and four-wire transformer connections is shown in Fig. 31 and a list of two-phase parallel combinations is shown in Fig. 32. These might or might not represent a certain change in the transformer leads. They are,

TWO-PHASE TRANSFORMER CONNECTIONS

however, only intended to represent those connections and combinations which can be made with the leads symmetrically located on the outside of the transformers.

For grounding two-phase systems several methods are employed, the best being those given in Fig. 33.

(*A*) represents a two-phase four-wire system, the two single-phases being independently operated, consequently two independent grounds are necessary.

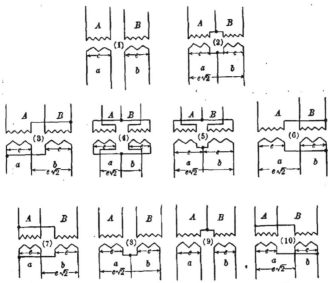

Fig. 31.—Complete list of two-phase transformer connections.

(*B*) also represents a two-phase four-wire system to be used for three-phase and two-phase at the same time. This is similar to the "*Taylor*" system excepting that two units instead of three are employed. The maximum voltage to ground is $a'-x$, or $E \times 0.866$.

(*C*) likewise represents a two-phase four-wire system with "T" three-phase primary. In this case the maximum voltage strain to ground is $E \sqrt{\frac{2}{3}}$ of voltage between terminals.

(*D*) represents a two-phase three-wire system. The maximum strain to ground in this case is full voltage of any phase.

38   STATIONARY TRANSFORMERS

(E) represents a "V" two unit system. The maximum voltage strain to ground is E 0.866, or the same as (B).

(F) represents the two-phase interconnected four- or five-wire

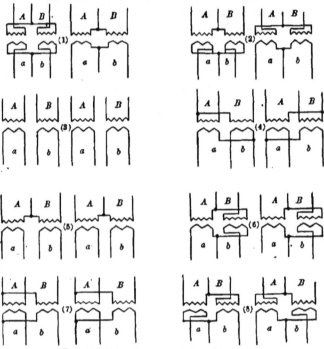

Fig. 32.—Two-phase parallel combinations.

system. The maximum voltage strain to ground is 50 per cent. of any phase voltage and the voltage across any two-phase terminals is 70.7 per cent. of full-phase voltage.

CHAPTER IV

THREE-PHASE TRANSFORMATION SYSTEM

General Principles.—In considering the question of three-phase transformation we have to deal with three alternating e.m.fs. and currents differing in phase by 120 degrees, as shown in Fig. 34.

One e.m.f. is represented by the line, $A B$, another by the line, $B C$, and the third by the line, $C A$. These three e.m.fs.

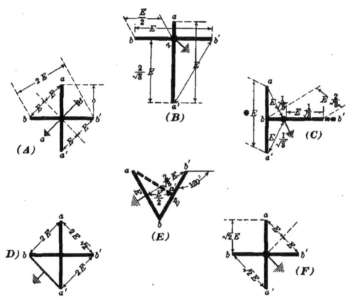

Fig. 33.—Methods of grounding two-phase systems.

and currents may be carried to three independent circuits requiring six wires, or a neutral wire, or common return wire may be used, where the three ends are joined together at x. The e.m.f. phase relations are represented diagrammatically by the lines, $a\, x$, $b\, x$, and $c\, x$; also, $A B$, $B C$, and $C A$. The arrows only indicate the positive directions in the mains and through the windings; this direction is chosen arbitrarily, therefore, it must be remembered that these arrows represent not the actual direction of the e.m.fs. or currents at any given instant, but merely the directions of the positive e.m.fs. or currents i.e., the positive direction through the circuit. Thus, in Fig. 32a the e.m.fs. or currents are considered positive when directed from the common junction x toward the ends, a b c.

In passing through the windings from a to b, which is the direction in which an e.m.f. must be generated to give an e.m.f. acting upon a receiving circuit from main a to main b, the winding, a,

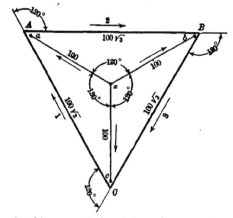

Fig. 34.—Graphic representation of three-phase currents and e.m.fs.

is passed through in a positive direction, and the winding, b, is passed through in a negative direction; similarly the e.m.f. from b to c, and the e.m.f. from main c to a. The e.m.f. between a and b is 30 degrees behind a in time-phase, and its effective value is

$$2 E \cos 30° = \sqrt{3}\ E; \qquad (14)$$

where E is the value of each of the e.m.fs. $a\ b$, and c.

With this connection the e.m.f. between any two leads, $a\ b$, $b\ c$, or $a\ c$, is equal to the e.m.f. in each winding, $a\ x$, $b\ x$, or $c\ x$ multiplied by the square root of three.

For current relations we see in Fig. 34 that a positive current in winding 1 produces a positive current in main A, and that a negative current in winding 2 produces a positive current in main A; therefore, the instantaneous value of the current in main

THREE-PHASE TRANSFORMATION SYSTEM 41

A is $I_1 - I_2$, where I_1 is the current in winding 1, and I_2 is the current in winding 2. Similarly, the instantaneous value of the current in main B is $I_2 - I_3$, and in main C, it is $I_3 - I_1$. The mean effective current in main A is 30 degrees behind I_1 in phase; and its effective value is the square root of three times the current in any of the different phases; so that with this connection the current in each main is the square root of three times the current in each winding.

When the three receiving circuits $a\ b\ c$ are equal in resistance and reactance, the three currents are equal; and each lags behind its e.m.f., $a\ b$, $b\ c$, and $a\ c$, by the same amount, and all are 120 time-degrees apart. The arrangement shown in Fig. 34, by the lines, a, b, and c, is called the "Y" or star connection of transformers. Each of these windings has one end connected to a neutral point, x; the three remaining ends, $a\ b\ c$, commonly called the receiving ends, are connected to the mains. The e.m.f. between the ends, or terminals of each receiving circuit is equal to $\sqrt{3}\ E$, where E is the e.m.f. between $a\ x$, $b\ x$, and $c\ x$. The current in each receiving circuit is equal to the current in the mains, $a\ b\ c$.

The resistance per phase cannot be measured directly between terminals, since there are two windings, or phases in series. Assuming that all the phases are alike, the resistance per phase is one-half the resistance between terminals. Should the resistances of the phases be equal, the resistance of any phase may be measured as follows:

The resistance between terminals $a\ b$ is:

$$\text{Resistance of } a\ b = R_3 + R_1.$$

The resistance between terminals $b\ c$ is:

$$\text{Resistance of } b\ c = R_1 + R_2.$$

The resistance between terminals $a\ c$ is:

$$\text{Resistance of } a\ c = R_3 + R_2.$$

Therefore:

$$R_3 = \frac{\text{Res. } a\ b - \text{Res. } b\ c + \text{Res. } a\ c}{2},$$

$$R_1 = \frac{\text{Res. } b\ c - \text{Res. } a\ c + \text{Res. } a\ b}{2}, \qquad (15)$$

$$R_2 = \frac{\text{Res. } a\ c - \text{Res. } a\ b + \text{Res. } b\ c}{2},$$

The method of connecting three-phase circuits shown in Fig. 34, where the windings, 1, 2 and 3, are connected in series at A, B and C, is called the delta connection. In this connection the e.m.f. on the receiving circuit is the same as that on the mains; and the current on each receiving circuit is equal to $\sqrt{3}$ times that in any winding, or $\sqrt{3}I$, where I is the current in 1, 2, or 3.

Assuming that all phases are alike, the resistance per phase is equal to the ratio of 3 to 2 times $\frac{3}{2}$ the resistance between A and B. In a delta connection there are two circuits between A and C, one through phase 1, and the other through phases 2 and 3 in series. From the law of divided circuits we have the joint resistance to two or more circuits in parallel is the reciprocal of the sum of the reciprocals of the resistances of the several branches. Hence, $\frac{3R}{2}$ is the resistance per phase R being the resistance of one winding with two others in parallel. The ohmic drop from terminal to terminal with current I in the line, is

$$\text{Ohmic drop} = \frac{3}{2} I \times R. \qquad (16)$$

To transform three-phase alternating current a number of different ways are employed. Several of the arrangements are:

1. Three single-phase transformers connected in star or in delta.

2. Two single-phase transformers connected in open-delta or in tee.

3. One three-phase transformer connected in star or in delta.

With three single-phase transformers the magnetic fluxes in the three transformers differ in phase by 120 time-degrees.

With two single-phase transformers the magnetic fluxes in them differ in time-phase by 120 degrees or by 90 degrees according to the connection employed.

With the three-phase transformer there are three magnetic fluxes differing in time-phase by 120 degrees.

The single-phase transformer weighs about 25 per cent. less than three separate transformers having the same total rating; its losses at full load are also about 25 per cent. less.

Two separate transformers, V-connected, weigh about the same as three single-phase transformers for the same power transmitted; the losses are also equal.

A three-phase transformer weighs about 16.5 per cent. less than three separate transformers; its losses are also about 16.5 per cent. less.

It is considered that for transformation of three-phase power the three-phase transformer is to be preferred to any other combination method.

In America it is customary to group together single-phase transformers for use on polyphase circuits, while in Europe the polyphase transformer is almost exclusively employed. However, a change is being made and more single-phase transformers are being employed in Europe than formerly. The relative merits of three-phase transformers and groups of three single-

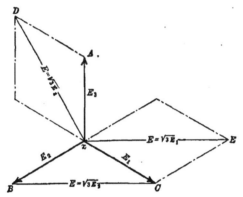

Fig. 35.—Graphic representation of three-phase e.m.fs.

phase transformers, in the transmission and distribution of power, form a question that is capable of being discussed from various standpoints.

A popular argument in favor of the three-phase transformer is the greater compactness of the transformer unit; and the favorite argument against the three-phase transformer is that if it becomes disabled all three sides of the system must be put out of service by disconnecting the apparatus for repair; whereas if a similar accident occurs to any one of three single-phase transformers in a delta-connected group, the removal of the defective transformer only affects one side of the system, and two-thirds of the total transformer capacity would go on working. The relative merits of the three-phase transformer and the combination of three single-phase transformers that may be employed for obtaining the same service are frequently discussed on the basis of the decrease in cost of the several types of transformers with increase in rating; and on such basis it has been shown that the three-phase transformer is the cheaper, while the other combinations are more expensive on account of requiring an equal or greater aggregate rating in smaller transformers. It should, however, be borne in mind that a three-phase transformer is not, generally speaking, so efficient as a single-phase transformer designed along the same lines and wound for the same total output.

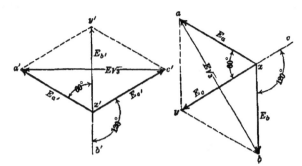

Fig. 36.—Vector sum of effective e.m.fs.

It is shown very clearly in Fig. 35 that a connection across the terminals, $B\ C$, $x\ E$, and $x\ D$, will receive a voltage which is the resultant of two e.m.fs. differing in time-phase by 120 degrees; or the result from adding the e.m.fs. of E_3 and E_1 at 60 degrees, which is equivalent to $E_2 \sqrt{3}$.

It is shown in Fig. 36, where one phase of a star-connected group of transformers is reversed, the resultant e.m.f., $x'\ y'$ and $x\ y$, is equivalent to the component e.m.fs., $x\ c'$, $x\ a'$ and $x\ b$, $x\ a$; consequently the resultant of all three e.m.fs. is zero. This is in accordance with Kirchhoff's law which states that where an alternating-current circuit branches, the effective current in the main circuit is the geometric, or vector sum of the effective currents in the separate branches. The modifications of the fundamental laws appertaining to this are discussed in many books treating alternating currents in theory, and the summary given here is for the purpose of comparison. However,

THREE-PHASE TRANSFORMATION SYSTEM 45

it is noted in Fig. 36 that the phase relations between a' and b', b' and c', a and c and b and c have been changed from 120 degrees to 60 degrees by reversing phases b' and c. The pressure between a and b and a' and c' is $E\sqrt{3}$, showing their phase relations to be unchanged.

For example, in Fig. 36, considering only secondary coils of three single-phase transformers that are supposed to be connected in star, but as a matter of fact connected as shown, let us assume that E is equal to 100 volts:

(a) What will be the voltage between terminals of E_a', E_c', and E_a, E_b, and E_c phases?

(b) What will be the voltage between terminals a' and c', and a and b?

(c) What will be the voltage across $a'\,y'$ and $y'\,c'$, also $a\,y$ and $y\,b$?

(a) The voltage acting on any of these windings is equal to $E\sqrt{3}$

$$100 \times \sqrt{3} = 173.2 \text{ volts.}$$

The voltage acting on E_a', E_b', etc., is

$$E_{a'} = \frac{1}{\sqrt{3}} \times 173.2 = 0.577 \times 173.2 = 100 \text{ volts.}$$

(b) The voltage between terminals a' and c', or a and b, is equal to $\sqrt{3}$ times the voltage acting on $E_{a'}$ etc., or

$$E\sqrt{3} = 100 \times 173.2 = 173.2 \text{ volts.}$$

(c) It is evident that the two e.m.fs., $a'\,x'$ and $c'\,x'$, are equivalent to their resultant, $y'\,x'$, which is equal and opposite to $b'\,x'$; the dotted lines, $a'\,y'$ and $y'\,c'$, are equal to E_a', E_b', etc., at 120 degrees apart; in other words,

$$E = a''\,y', \text{ etc.}, = \frac{1}{\sqrt{3}} \times 173.2$$

$$= 0.577 \times 173.2 = 100 \text{ volts.}$$

In Fig. 37 is shown a diagram of three-phase currents in which x, y, z are equal and 120 degrees apart, the currents in the leads, a, b, c, are 120 degrees apart and each equal to $\sqrt{3}$ times the current in each of the three circuits X, Y and Z.

STATIONARY TRANSFORMERS

The current in each lead is shown made up of two equal components which are 60 degrees apart.

As an example showing the use of Fig. 35, assume a circuit, of three single-phase transformers delta-connected. What will be the

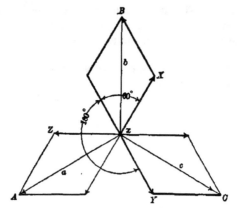

Fig. 37.—Geometric sum of three-phase currents.

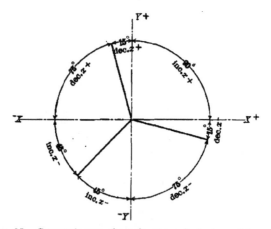

Fig. 38.—Geometric sum of e.m.fs. at any instant equal to zero.

current through X-phase winding if the current in b lead is 500 amperes?

Since one component, X, has in the lead the same sign as it has

THREE-PHASE TRANSFORMATION SYSTEM

in its own transformer winding, and the other component, Y, has in the lead the opposite sign to that which it has in its own transformer winding, X, is represented in lead by the same vector as in its own transformer winding; while the other component, Y, is represented in the lead by a vector 180 degrees from that which represents it in its own circuit, therefore, $X = \frac{1}{\sqrt{3}} 500$.

The instantaneous values of the currents in any one wire of a three-phase system are equal and opposite to the algebraic sum of the currents in the other two sides. Therefore, the algebraic sum of e.m.fs. or currents at any instant is equal to zero. This fact is shown in Fig. 38, the geometric sum of the three lines, A, B, C, being equal to zero at any instant.

Single-phase transformers may be connected to a three-phase system in any of the following methods:

Delta-star group with a delta-star.
Star-star group with a star-star.
Delta-delta group with a delta-delta.
Star-delta group with a star-delta.
Delta-star group with a star-delta.
Delta-delta group with a star-star.

It must, however, be remembered that it is impossible to connect the following combinations, because the displacement of phases which occurs, when an attempt is made to connect the secondaries, will result in a partial short-circuit.

Delta-star with a star-star.
Delta-delta with a star-delta.
Delta-star with a delta-delta.
Star-delta with a star-star.

As is well known there are four ways in which three single-phase transformers may be connected between primary and secondary three-phase circuits. The arrangements may be described as the delta-delta, star-star, star-delta, and delta-star.

In winding transformers for high voltage the star connection has the advantage of reducing the voltage on an individual unit, thus permitting a reduction in the number of turns and an increase in the size of the conductors, making the coils easier to wind and easier to insulate. The delta-delta connection nevertheless has a probable advantage over the star-delta arrangement, in that if one transformer of a group of three should become disabled, the two remaining ones will continue to deliver three-phase currents with a capacity equal to approximately two-thirds of the original output of the group. Fig. 39 shows a delta-delta arrangement. The e.m.f. between the mains is the same as that in any one transformer measured between terminals. The current in the line is $\sqrt{3}$ times that in any one transformer winding.

Each transformer must be wound for the full-line voltage and for 57.7 per cent. line current. The greater number of turns in the winding, together with the insulation between turns, necessitate a larger and more expensive coil than the star connection.

For another reason the delta-delta connection may be preferable

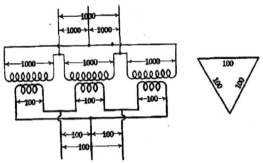

Fig. 39.—Delta-delta connection of transformers.

to the star, inasmuch as the arrangement is not affected even though one transformer may be entirely disconnected, in which case it is practically assumed that the two remaining transformers have exactly a carrying capacity of 85 per cent. of $\frac{2}{3} = 0.567$.

In a delta connected group of transformers the current in each phase winding is $\frac{I}{\sqrt{3}}$, I being the line current, and if a phase displacement exists, the total power for the three phases will be
$$\sqrt{3} \times E \times I \times \cos \theta. \tag{17}$$

Assume the voltage between any two mains of a three-phase system to be 1000 volts; the current in the mains, 100 amperes, and the angle of time-lag, 45 degrees. What will be the e.m.f. acting on each phase, the current in each phase, and the output?

The e.m.f. on each phase of a delta-connected group of transformers is the same as that across the terminals of any one trans-

THREE-PHASE TRANSFORMATION SYSTEM 49

former. The line current is $\sqrt{3}$ times that in each transformer winding, or $\sqrt{3} \times 57.7 = 100$ amperes, therefore the current in each phase is $100 \times \dfrac{1}{\sqrt{3}} = 57.7$ amperes. The output being $\sqrt{3}\,E\,I\,$- Cos $\theta =$

$$1.732 \times 1{,}000 \times 100 \times .71 = 123 \text{ kw. approx.}$$

In the star-star arrangement each transformer has one terminal connected to a common junction, or neutral point; the three remaining ends are connected to the three-phase mains.

The number of turns in a transformer winding for star connection is 57.7 per cent. of that required for delta connection and the

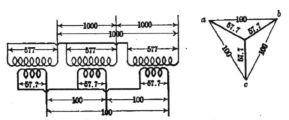

Fig. 40.—Star-star connection of transformers.

cross-section of the conductors must be correspondingly greater for the same output. The star connection requires the use of three transformers, and if anything goes wrong with one of them, the whole group *might* become disabled.

The arrangement shown in Fig. 40 is known as the "star" or "Y" system, and is especially convenient and economical in distributing systems, in that a fourth wire may be led from the neutral point of the three secondaries.

The voltage between the neutral point and any one of the outside wires is $\dfrac{1}{\sqrt{3}}$ of the voltage between the outside wires, namely

$$1000 \times \dfrac{1}{\sqrt{3}} = 1{,}000 \times 0.577 = 577 \text{ volts.}$$

The current in each phase of a star-connected group of transformers is the same as that in the mains.

Fig. 41 shows a star-star connection in which one of the

secondary windings is reversed. It may be noted that the phase relations of phase *c* have changed the relations of *a c* and *b c* from 120 degrees to 60 degrees by the reversal of one transformer connection. The resultant e.m.fs., *a c* and *b c*, are each $\dfrac{1}{\sqrt{3}} =$

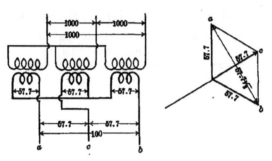

Fig. 41.—Star-star connection of transformers, one-phase reversed.

57.7, but in reality should be 100 volts—the voltage between *a* and *b* is $57.7 \times \sqrt{3} = 100$. In star-connecting three single-phase transformers it is quite possible to have one of the transformers reversed as shown.

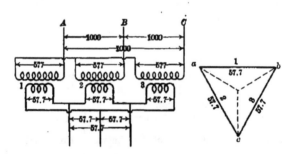

Fig. 42.—Star-delta connections of transformers.

In the star-delta arrangement shown in Fig. 42 the ratio of transformation is $\dfrac{1}{\sqrt{3}}$, or 0.577 times the ratio of secondary to primary turns, and the e.m.f. acting on each secondary circuit is the same as that between the mains.

THREE-PHASE TRANSFORMATION SYSTEM

For example, let us assume that 1000 volts are impressed on the primary mains. The voltage between any two secondary mains is the same as that generated in each transformer, namely, 57.7 volts, or $100 \times \dfrac{1}{\sqrt{3}} = 57.7$ volts.

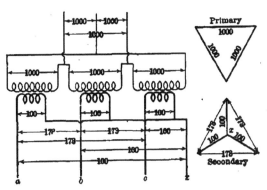

FIG. 43.—Delta-star four-wire connection of transformers.

Fig. 41 shows a delta-star connection using three single-phase transformers. From the neutral point of the secondary star connection, a wire may be brought out, serving a purpose similar to that of the neutral wire in the three-wire Edison system, it

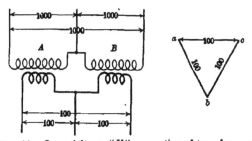

FIG. 44.—Open-delta or "V" connection of transformers.

being without current when the load is balanced. The ratio of transformation for the delta-star arrangement is $\sqrt{3}$, or 1.732 times the ratio of secondary to primary turns.

The advantage of this secondary connection lies in the fact

that each transformer need be wound for only 57.7 per cent., of the voltage on the mains.

In the arrangement, commonly called "V" or "open-delta" the voltage across the open ends of two transformers is the resultant of the voltages of the other two phases; see Fig. 44. This method requires about 16 per cent. more transformer capacity than any of the previous three-phase transformations shown, assuming the same efficiency of transformation, heating, and total power transformed.

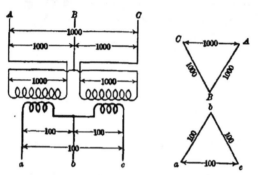

FIG. 45.—"V" connection of transformers with secondary windings connected in opposite directions.

In comparing the kv-a capacity of two single-phase transformers connected in open delta, with three similar transformers connected in delta, the kv-a capacity is approximately $1 \div \sqrt{3} =$ 58 per cent. This is due to the fact that for the delta connection the current and voltage of each transformer are in phase with each other, while for the open-delta connection the current and voltage of each transformer are 30 degrees out of phase with each other.

With the open-delta method a slight unbalancing may exist, due to the different impedances in the middle main and the two outside mains, the impedance in the middle main being the algebraic sum of the impedances in the two outside mains.

The open-delta arrangement, (where the primary is connected like that shown in Fig. 45) is in every respect equivalent to the open-delta connection represented in Fig. 42. The primaries are connected in a reverse direction, or 180 degrees apart in phase with the secondary.

THREE-PHASE TRANSFORMATION SYSTEM

The vector diagram shows the changed phase relations of primary to secondary. By connecting two single-phase transformers in the opposite direction, that is to say, connecting the secondary like that of the primary, and the primary like that of the secondary, we obtain the same transformation characteristics.

Like that of the open-delta arrangement the tee method requires only two single-phase transformers.

As regards the cost of equipment and the efficiency in operation the tee arrangement is equal to either the open-delta, the star or the delta methods.

The tee arrangement is represented in Fig. 46. The end of one transformer is connected to the middle of the other.

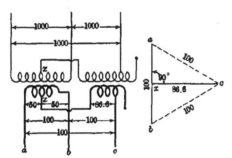

Fig. 46.—Tee or "T" connection of transformers.

The number of turns on ab is $\dfrac{2}{\sqrt{3}} = 1.16$ times the number of turns on xc. Its ability to maintain balanced phase relations is no better than the open-delta arrangement, and in no case is it preferable to either the star or delta methods of connecting three single-phase transformers.

It is worthy of note that the transformer which has one end tapped to the middle of the other transformer need not be designed for exactly $\dfrac{\sqrt{3}}{2} = 86.6$ per cent. of the voltage; the normal voltage of one can be 90 per cent. of the other, without producing detrimental results.

Another three-phase combination is shown in Fig. 47. In this case it is assumed that the three single-phase transformers were originally connected in star at oA, oB and oC as shown by dotted line in the vector diagram.

To change the transformation to the combined system of the star-delta, the end of winding oB is connected with the point o' of the phase oA, and the end O of the phase oC with the point o'' of the phase oB so that the vector oB becomes $o'B'$ and oC becomes $o''C'$. The length of the three vectors is proportional to the number of coils per phase, all coils having the same num-

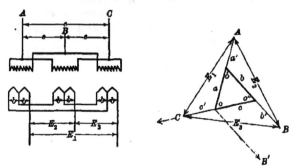

FIG. 47.—Delta connection with 50 per cent. of winding reversed to obtain another phase relation for maximum voltage.

ber of turns. The end o of phase oA is connected to the point of the phase $o''C'$.

Going over the list of common three-phase connections we find there are only four delta-delta systems in common use, these being shown in Fig. 48.

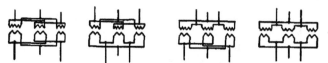

FIG. 48.—Common delta connections.

These, and a combination of them may be so arranged as to give not less than sixteen different changes; all these changes being given in Fig. 49.

Taking away one transformer of each of the above combinations we find open-delta connections like those given in Fig. 50.

On carefully looking over the eighteen different delta-delta combinations given in the above figures, it will be noted that the first ten represent changes of one or two of the transformers while

THREE-PHASE TRANSFORMATION SYSTEM

the remaining eight have all their secondaries connected in a direction opposite to their respective primary windings.

There are also four common star-star connections in general use, these being shown in Fig. 51.

Like the delta-delta, these four star-star connections can be arranged to give sixteen different combinations as will be seen from Fig. 52.

Continuing further through the series of three-phase connections we come to delta-star or star-delta combinations. The star-delta and delta-star in common use are given in Fig. 53.

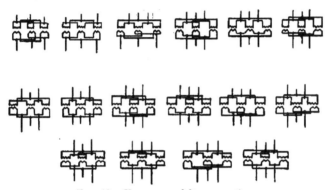

Fig. 49.—Uncommon delta connections.

The uncommon systems of star-delta transformer connections are shown in Fig. 54. These may be changed about to delta-star as desired, thus making twice the number. It must be remembered, however, that none of them can be changed about from one set of combinations to another.

POLARITY IN THREE-PHASE TRANSFORMER CONNECTIONS

It is customary for all transformer manufacturers to arrange transformer windings so that left- and right-hand primary and secondary leads are located the same for all classes and types. It is not, however, customary for operating companies and others to do so after a transformer has been pulled to pieces. The majority seems to follow manufacturer's prints and wiring diagrams of transformer connections coming from the factory, all of which are assumed to have the same relative direction of primary and secondary windings, thus permitting similarly located leads leaving the transformer tank to be connected together for parallel operation if desired without the necessity of finding the polarity.

Large and moderate-sized high-voltage transformers are rarely used for single-phase service. They may be of single or three-phase design but they are used almost exclusively on three-phase systems, and connected in star or delta or a combination of the

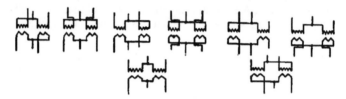

Fig. 50.—Uncommon open-delta connections.

two methods. Single-phase units in large and medium and small sizes (6000 to 0.5 kv-a) are very common. They afford a better opportunity for different polarities than three-phase units, which are also made in large and medium and small sizes (14,000 to 1.0 kv-a).

The delta and star methods of connections generally referred to as "conventional" are understood to have all the primary and secondary windings wound in opposite directions around

Fig. 51.—Common star connections.

the core. This conventional method is not always followed as problems arise in every-day practice showing that it is not always carried out.

Three single-phase transformers or one three-phase transformer may have their windings arranged in the following manner:
  (1) The three primary and secondary windings wound in the same direction—*positive direction.*
  (2) The three primary and secondary windings wound in the same direction—*negative direction.*

THREE-PHASE TRANSFORMATION SYSTEM

(3) The three primary windings wound in opposite direction to the secondary windings—*positive direction*.
(4) The three primary windings wound in opposite direction to the secondary windings—*negative direction*.
(5) One primary winding of the group of three may be wound opposite to the remaining two primaries and the three secondaries—*positive direction*.

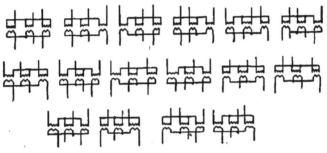

Fig. 52.—Uncommon star connections

(6) One primary winding of the group of three may be wound opposite to any of the remaining two primaries and the three secondaries—*negative direction*.
(7) Two primary windings of the group of three may be wound opposite to the one remaining primary and three secondary windings—*positive direction*.

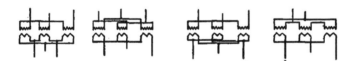

Fig. 53.—Common star-delta connections.

(8) Two primary windings of the group of three may be wound opposite to the one remaining primary and the three secondary windings—*negative direction*.
(9) One unit may have its primary and secondary windings different to the remaining two remaining units, as:
   (a) The unit on the left—*positive direction*.
   (b) The center unit—*positive direction*.
   (c) The unit on the right—*positive direction*.

58   STATIONARY TRANSFORMERS

(10) One unit may have its primary but not the secondary windings different to the remaining two units, as (9) but in the *negative direction*.

(11) Two units may have their primary and secondary windings different to the remaining unit, as:
  (a) The left and center units—*positive direction*.
  (b) The right and center units—*positive direction*.
  (c) The left and right units—*positive direction*.

(12) Two units may have their primary but not secondary windings different to the one remaining unit, as (11) but in the *negative direction*.

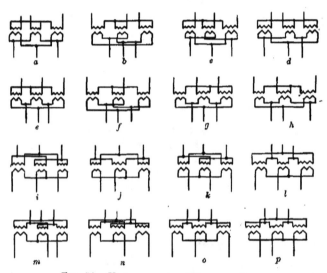

Fig. 54.—Uncommon star-delta connections.

Transformers of the same rating and of the same make are as a rule assumed to have the same polarity, impedance and ratio of transformation. In all cases where transformers are not of the same make it is advisable that the secondary connections be subjected to polarity tests before connecting them in delta or star. The mistakes which can be made, are:

(1) One or more reverse windings.
(2) Internal leads crossed.
(3) Where transformers are located some distance apart,

THREE-PHASE TRANSFORMATION SYSTEM

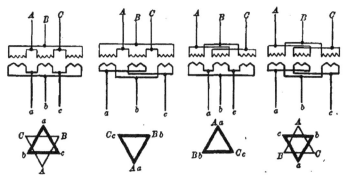

Fig. 55.—Three-phase polarity combinations.

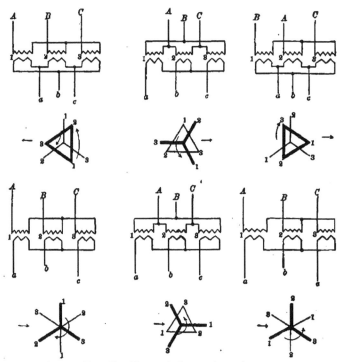

Fig. 56.—Phase relations and rotation.

STATIONARY TRANSFORMERS

and where connections are made at some distance, the leads may become crossed.

The reversal of two leads of either the primary or secondary will *reverse the polarity*, this being (so far as the external connections are concerned) the same as reversing one winding.

Reversing the line leads of a star-star, delta-delta, star-delta or delta-star will not reverse the polarity since the transformer leads themselves must be changed in order to make the change in polarity. This should be particularly noted when parallel

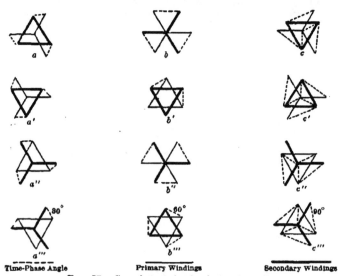

Fig. 57.—Complete cycle of polarity changes.

operation is desired, for, though the phase relations of two transformer groups may be the same parallel operation might be impossible (see Fig. 55).

The effect of a reversal of one or two primary windings is not the same for the four systems of star and delta, but is:

For Star-star: The reversal of one or two primary windings will not produce a short-circuit, but will produce a difference in phase relations and voltages. The maximum voltage will not be greater than the line voltage (see Fig. 58).

For Delta-delta: The reversal of one or two primary windings will immediately produce a short-circuit when the secondary delta is closed. The maximum voltage difference will be $2E$ or double-line voltage (see Figs. 59 and 60).

For Star-delta: The reversal of one or two primary windings will immediately produce a short-circuit when the secondary delta is closed. The maximum voltage difference at that point which closes the delta will be $2E$ or double-line voltage.

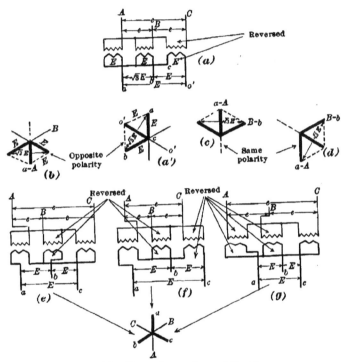

Fig. 58.—Three-phase polarity difficulties.

For Delta-star: The reversal of one or two primary windings will not produce a short-circuit on the secondary side when the star is made. The maximum voltage difference will be E or line voltage, but voltages and phase relations will be unequal.

The thick, black lines shown in Figs. 55, 56, 57, 58, 59 and 60 represent secondary windings.

Fig. 57 gives all the possible star and delta polarity combinations, or, sixteen as already explained.

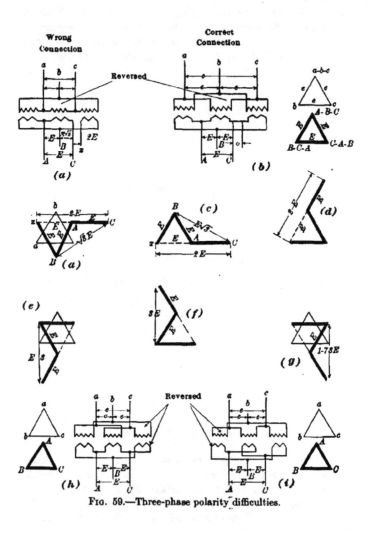

Fig. 59.—Three-phase polarity difficulties.

THREE-PHASE TRANSFORMATION SYSTEM

Fig. 58 shows the effects of a reversed winding or transformer. Unlike the delta-delta or star-delta no short-circuit exists when secondary connections are completed.

Fig. 59 (a) and (b) shows wrong and right method of connecting the delta secondary when one winding is reversed. Across C_x of (a) double-line voltage is obtained thus making the delta secondary impossible and only operative when connected as

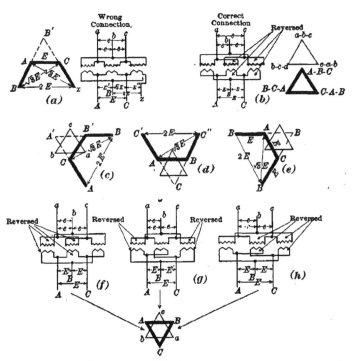

FIG. 60.—Three-phase polarity difficulties with two transformers reversed.

shown at (b). Both (a) and (b) have their primary and secondary windings wound in opposite directions, as also (h) and (i).

Fig. 60 (a) and (b) also show a correct and incorrect way of connecting the windings in delta when two transformers of a group of three are reversed. As before, double voltage or $2E$ is the maximum difference in voltage which can exist. Both (a) and (b) and (f), (g) and (h) are assumed to have their primary and

64   STATIONARY TRANSFORMERS

secondary windings of the same polarity. The primaries and secondaries of (c) and (d) are assumed to have opposite polarity.

Fig. 57 may also have parallel opearting polarity combinations represented as a complete cycle of changes.

From the above it is evident that from the four systems of delta-delta, delta-star, star-star and star-delta, we obtain 48 different combinations.

PARALLEL CONNECTIONS

After we have gotten through with these various combinations our next difficulty is in knowing just which one combination can be operated in parallel with another. With the primary and secondary windings arranged as is usual in practice, not less than eight different parallel combinations may be made (although sixteen are possible) these being given in Fig. 61. Other interesting series of parallel combinations are given in Figs. 62, 63, 64 and 65.

The problem of paralleling systems, no matter whether they be generating or transforming, resolve themselves into such factors as equal frequency, equal voltages which must be in time-phase, etc. The manufacturer is responsible for a fair part of the troubles of operating engineers from the apparatus point of

THREE-PHASE TRANSFORMATION SYSTEM 65

view, but the engineer himself shoulders the major portion of responsibility so far as the operating life of the apparatus is concerned whether the apparatus be operated singly or in multiple. The manufacturer can, and endeavors, to design apparatus

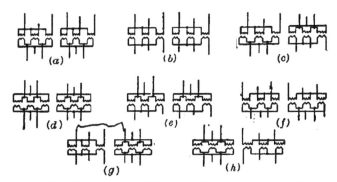

FIG. 61.—Three-phase parallel combinations in common practice.

and deliver same to the customer such that identical polarity, equal capacity and equal voltages are obtained and satisfactory parallel operation made possible. However, one particular manufacturer cannot be held responsible for the methods, etc.,

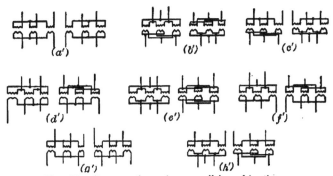

FIG. 62.—Common three-phase parallel combinations.

of another manufacturer who may design the same size and class of apparatus which differs in ratio of high-voltage to low-voltage turns, in impedance and also in its polarity.

The fault and responsibility, generally speaking, rests with the engineers themselves. Large and moderate size transformers are most always operated in parallel and consequently when ordering other transformers for the purpose of parallel operation, and when the order is from a different maker, certain specifications should be covered if the delta circulating currents are to be avoided and the apparatus is to be satisfactorily operated in parallel.

In the parallel operation of delta and star systems, two main factors must be kept in mind after due consideration has been given to the design of the apparatus to be placed in parallel. These are:

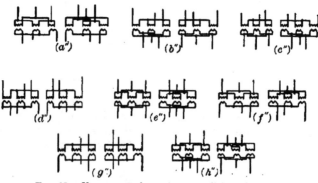

FIG. 63.—Uncommon three-phase parallel combinations.

(a) No condition is possible whereby apparatus connected in delta on both the high-voltage and low-voltage sides can be made to parallel with another piece of apparatus connected either in delta on the high-voltage side and star on the low-voltage side or in star on the high-voltage side and delta on the low-voltage side. However, a condition is possible whereby apparatus connected in delta on the high-voltage side and star on the low-voltage side can be made to parallel with *several* combinations of another piece of apparatus connected in star on the high-voltage side and delta on the low-voltage side.

(b) Some combinations of one group of apparatus or one polyphase unit connected exactly the same cannot be made to operate in parallel, as, for instance, a delta-star may have a combination that will not parallel with another polyphase unit or group of apparatus connected delta-star.

As already stated, to secure perfect parallel operation the two sets of apparatus should have exactly the same ratio of transformation, the same IR drop and the same impedance drop. Too little notice appears to be taken by operating engineers of the right size and type, identical characteristics, and the ability of transformers to share equal loads when operated in parallel.

Even though these conditions are obtained it will not follow that parallel operation is secured nor possible. For instance, assuming conditions are such that the characteristics of two sets of apparatus are identical and that they have exactly the same

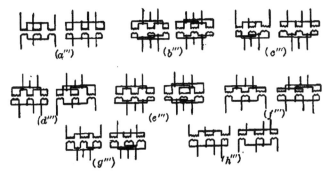

Fig. 64.—Uncommon three-phase parallel combinations.

ratio of high-voltage to low-voltage turns, the same IR drop and the same impedance drop, it will not follow that by using identical connections at the case of the apparatus, that a star-star can be operated in parallel with another star-star or delta-delta.

If the designs of two polyphase units are such that satisfactory operation may be carried out, the next step for the operator is to ascertain their phase relations; that is to say, see if their phase relations between the high voltage and low voltage are identical, for, apparatus coming from the factory may or may not have the same polarity. If two three-phase transformer groups have in themselves or between them a difference in polarity (positive in one and negative in another) it will be impossible to operate them in parallel when arranging their external connections symmetrically.

For the satisfactory parallel operation of three-phase systems it is necessary first to ascertain that:

(a) Each single-phase unit or each phase of the polyphase unit has the same ratio of transformation, the same IR drop and the same impedance drop.

(b) The phase relation is the same (see Fig. 56).

(c) The polarity is the same (see Fig. 55).

With a slight difference in ratio, unbalanced secondary voltages or a circulating current will result.

With a difference in impedance the total efficiency may become badly affected, though not so pronounced as that which would exist if each unit or phase was tied directly together and afterward connected in star or delta.

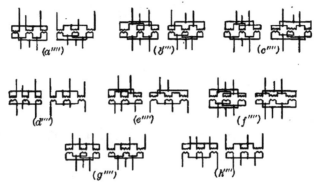

FIG. 65.—Uncommon three-phase parallel combinations.

Phase rotations sometimes offers complications where mixed systems of delta and star are employed. To reverse phase rotation two lines (not transformer terminal leads) must be reversed.

Polarity complications are even worse and sometimes present much difficulty. To reverse the polarity two transformer leads (not line leads) must be reversed (see Figs. 58, 59 and 60).

Ordinary parallel operation and ordinary connections for parallel operation are quite simple. Complications set in when two or more groups of different system connections have to be operated in parallel. On some of our present day large and centralized systems, it happens that certain "make-shift" parallel combinations must be made with the available apparatus, perhaps in the stations themselves or on some part of the general distribution system. Usually the chief operating engineer has a list of available transformers at each station and center which

THREE-PHASE TRANSFORMATION SYSTEM 69

facilitate their adoption when urgently needed. On the occurrence of a breakdown in any station or center he will issue an order to make up a temporary substitute of transformer or transformers as the requirement calls for. This may mean a simple change and use of ordinary connections, or it may mean a substitute of one or more transformers of odd voltages or kw. capacity or both. In fact it may mean, in order to deliver the amount of energy necessary and continue operation, that he will have to resort to connecting certain transformers in series and others in multiple series, etc. It may also mean that such unusual parallel combinations as those shown in Figs. 66 and 66a will have to be made.

A number of delta-star and delta-delta parallel combinations similar to above might be made but great care must be taken to phase out the secondary windings of each group before tying them together, for, as in well known, a straight delta-star cannot

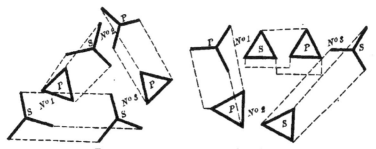

Fig. 66.—Unusual parallel combinations.

be tied in with a delta-delta and *vice versa*, nor a star-star with a delta-star although in a somewhat modified form this has been done in Fig. 66. With these and similar "make-shift" combinations of odd voltages and capacities engineers encounter from time to time on our larger systems, the essential points to remember are the difference in phase position and rotation and also the different impedences of the various transformers differing very much in sizes. The two former difficulties show themselves immediately parallel operation is tried but the latter only demonstrates itself by excessive heating or a burn-out and consequently is a serious point and one very often neglected by operating engineers in their rush to get the system in regular working order.

STATIONARY TRANSFORMERS

When refering to parallel operation it is oftentimes stated that two or more three-phase groups or two or more three-phase transformers may be connected in parallel on the low- or high-voltage side and yet it may not be possible to connect them together on the other side. This condition does not constitute parallel operation of transformers. Parallel operation of transformers is always understood to mean that before a condition of parallel operation can exist both the primaries and the secondaries respectively must be tied together, neither the one nor the other alone constituting parallel operation.

CHOICE OF CONNECTIONS

In the connection of power transformers for transmission systems there is a choice between four *main* combinations for three-phase and three-phase two-phase systems, namely, delta and star, see Fig. 67.

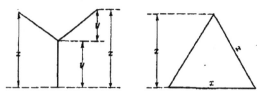

Fig. 67.—The two common systems.

Where $x = y\sqrt{3}$, or 100 per cent.

and $y = x\dfrac{1}{\sqrt{3}}$ or nearly 57.7 per cent. of full voltage between lines in case of the star connection;

or

$$100.0 \div \sqrt{3} = 57.7 \text{ per cent.}$$

The delta connection, where $x = 100$ per cent., the voltage is that shown between lines; or $577.7 \times \sqrt{3} = 100$ per cent.
and three-phase to two-phase; namely, two-transformer "T" and three-transformer "T", see Fig. 68.
Where

$$y = \frac{\sqrt{3}}{2} = 86.6 \text{ per cent. of the voltage between transformer terminals } a', b' \text{ and } c.$$

$x = y \dfrac{2}{\sqrt{3}} 100$ per cent., or full voltage between transformer terminals a', b' and c.

$z = $ Full terminal voltage a', b' and c, corresponding to a ratio of 1.0 to 1.15 of abc values.

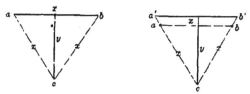

Fig. 68.—The uncommon systems.

Star vs. **Delta.**—It is shown in Fig. 69 that should one of the three single-phase transformers be cut out of star, or one of the leads joined to the neutral point be disconnected, there will exist only one voltage instead of three, across the three different phases.

This disadvantage is detrimental to three-phase working of the star arrangement, inasmuch as two equal and normal phase

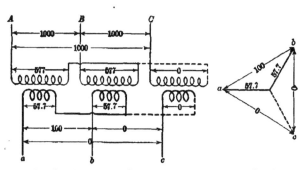

Fig. 69.—Result of one transformer of a star-connected primary and secondary group being cut out of circuit.

voltages of the three-phase system are disabled, leaving one phase voltage which may be distorted to some degree, depending on conditions.

On the other hand, should one phase, or one transformer of a delta-connected group be disconnected from the remaining two, as shown in Fig. 70, there will exist the same voltage between the three different phases, and practically the same operating conditions.

The result obtained by cutting out of delta one transformer, is simply the introduction of open delta, which has a rating of a little over one-half the total capacity; or more correctly, the rating of transformer capacity is

$$85 \text{ per cent.} \times 0.6666 = 0.5665$$

of three transformers of the same size connected in delta.

In the past it has frequently been urged against the use of three-phase transformers with interlinked magnetic circuits that if one or more windings become disabled by grounding, short-

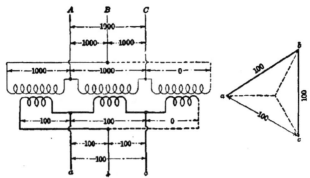

Fig. 70.—Result of a delta-connected group of transformers with one transformer disconnected.

circuiting, or through any other defect, it is impossible to operate to any degree of satisfaction from the two undamaged windings of the other phases, as would be the case if a single-phase transformer were used in each phase of the polyphase system.

All that is necessary is to short-circuit the primary and secondary windings of the damaged transformer upon itself, as shown in Fig. 71. The windings thus short-circuited will choke down the flux passed through the portion of the core surrounded by them, without producing in any portion of the winding a current greater than a small fraction of the current which normally exists at full load.

With one phase short-circuited on itself as mentioned above, the two remaining phases may be reconnected in open delta in tee or in star-delta for transforming from three-phase to three-phase; or the windings may be connected in series or parallel for single-phase transformation. This method of getting over a trouble is only applicable where transformers are of the shell type.

The relative advantages of the delta-delta and delta-star systems are still, and will always be disputed and wide open for discussion. They possess the following advantages and disadvantages, respectively:

Delta-delta (non-grounded). When one phase is cut out the remaining two phases can be made to deliver approximately 58

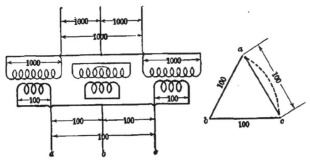

Fig. 71.—Result of operating a delta-connected transformer with one winding disabled and short-circuited on itself.

per cent. of the full load rating of transformer (in the case of a three-phase shell type) or three single-phase transformers.

Delta-star (neutral grounded). Advantage of reducing the cost of high voltage line insulators for equal line voltage, which is a very large item when dealing with long-distance lines; their size need only be approximately 58 per cent. of that used on a line using the delta system.

It is also possible, under certain conditions, to operate and deliver three-phase currents when one phase or one line conductor is on the ground.

The disadvantages are:

Delta-delta (non-grounded). Larger transformer or transformers and larger line insulators for the same line voltage.

Delta-star (neutral grounded). Not always in a position to operate when one phase or one line conductor is cut out.

Table I gives a comparison of the four common three-phase systems.

TABLE I

Governing factors	Star-delta to delta-star	Star-star to any	Delta-delta to any	Delta-star to star-delta
Cheapest cost....	Third.	First.	Fourth.	Second.
Best operated....	Third.	Fourth.	First.	Second.
Least potential strain.	Fourth.	Third.	Second.	First.

Cheapest Cost.—This represents the lowest price for transformers and system (complete) of equal kw. capacity and terminal line voltage.

Best Operated.—All the star connections are assumed to have their neutral points grounded, and the generators in each case star-connected and grounded. Each case is figured to have either one line on the ground or one transformer or one phase disabled by a burnt-out unit or other fault.

Least Potential Strain.—This represents the worst voltage strain that can be placed on the line and receiving station transformers, no matter what changes of phase relation might occur as the result of open connections, short-circuits or any combination of these.

It is quite evident that the delta-star to star-delta system is the best all round system to have. It will also be noted that this system takes a second place of importance in the "best operated" list, for the reason that a ground on one line short-circuits that phase whereas with the delta-delta to delta-delta (assuming all other phases thoroughly insulated which is almost a practical impossibility on high voltage systems) the system is not interrupted. Although a doubtful question it is placed in its favor, but beyond this weak point the delta-star to star-delta is equally as good and reliable as the delta-delta to delta-delta and about equally able to operate and furnish three-phase currents with only two transformers. The right order of importance giving the best system is:

First. Delta-star to star-delta.
Second. Delta-delta to delta-delta.
Third. Star-delta to any combination.
Fourth. Star-star to any combination.

Depending on the voltage and size of transformers the relative cost will vary, but the advantages given are about correct for almost all high-voltage systems. The same thing applies in the case of the best system for operation, but for general cases Table I list will be found to be close. In fact, if such questions as the third harmonics and the resulting flow of unbalanced currents in the closed delta-delta to delta-delta be considered, there is still something better in favor of the delta-star to star-delta system.

An advantage somewhat in doubt is, that only one high-voltage terminal of a delta-star to star-delta system is subjected to the full incoming high-voltage surges, whereas the delta-delta system has always two or double the number of high-voltage transformer terminals connected to the transmission line, and, of course, almost double the chance of trouble due to high-voltage surges. Impulses coming in over a line will enter the high-voltage windings of transformers from both ends, and even though the effect be in some degree divided if, say, two lines are disturbed, it will not have the same total factor of safety as would the delta-star to star-delta system, which has the uninterrupted facility of dividing the impulse between two transformer windings. The same impulse will also divide its effect between two or three transformer windings of a delta-delta system but not with the same effect because of the connections.

In case one line terminal is disturbed by an incoming surge only one transformer winding of a delta-star to star-delta system is effected. The disturbance will, of course, be greater than that affecting a delta-delta system, but the difference will not generally be great enough to cause a break-down on one system and not on the other; in fact two transformer windings of a delta-delta system may, in the majority of cases, be injured to one of the delta-star to star-delta systems.

Another very important advantage, particularly so where very high voltages are employed, not in favor of either the delta-delta or star-delta (delta on the high-voltage side) is that each transformer of a group of three has its winding terminals exposed to every line surge and consequently double the possibility of trouble that can occur on delta-star to star-delta systems (star on the high-voltage line side); also, the coils of a delta-delta or star-delta (delta on the high-voltage side) have a greater number of turns of smaller cross-section for a given kw. capacity and consequently are more liable to mechanical failure than a delta-star connected system (star on the high-voltage side).

A further disadvantage of the high-voltage delta is that if it is thought necessary to ground the delta as a safeguard for high-voltage stresses, it will require a group of transformers with interconnected phases or a star-delta connection; therefore additional apparatus is required whereas the delta-star system can be grounded direct without any additional expense.

As regards some of the advantages of switching, the delta-star has a further advantage over the delta-delta. The delta-star (star on the high-voltage side or low-voltage side) can give claim to an advantage by its simple, effective and cheap arrangement of switches in all stations where it is found necessary to use air-break disconnecting or single-pole switches of any kind installed on each high- and low-voltage lead. Only two such switches instead of four per transformer are needed, the remaining two leads being solidly connected to a neutral bus-bar grounded direct or through a resistance as thought desirable. Its advantage in this respect is important on very high-voltage systems where stations are cramped for space; as an illustration of this take one three-phase group and we find:

> Twelve switches are required for a *delta-delta* group of three single-phase transformers.
> Nine switches are required for a *delta-star* group of three single-phase transformers.
> Six switches are required for a *star-star* group of three single-phase transformers.

In this it is also well to remember that a spare single-phase transformer arranged to replace any of the three single-phase transformers of the group, will require the same number of switches, or twelve, nine and six respectively. All the switching referred to here only holds good when the neutral point of the star is grounded.

There exist a large variety of system connections quite different from the common ones mentioned above. For instance, it is not unusual now to see in one station a group of three single-phase transformers operating in parallel with two single-phase transformers, the two groups being connected in delta and open-delta respectively. It is well known that the open-delta system does not claim to possess any merits over any of the common systems above mentioned, but it is oftentimes necessary to fall back

THREE-PHASE TRANSFORMATION SYSTEM 77

on this system as a stand-by or in an emergency and from this point of view it becomes very useful. Now, in certain cases of station wiring layout, particularly in those stations operating above 50,000 volts, it might be only possible to use the delta connection for parallel operation after considerable loss of time rearranging the wiring. The wiring layout of a station, however, might be such with respect to the location of transformers that with a disabled unit of a group of three connected in delta-delta it would not be possible to connect the two remaining transformers in parallel and in open-delta with a group of three others

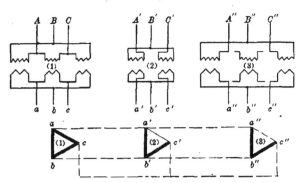

Fig. 72.—Correct method of connecting the transformers.

located some distance away. Suppose for example that the station bus-bar wiring in the high- and low-tension bus-bar compartments is arranged to meet any delta and open-delta combination and that two groups of delta and one group of open delta-connected transformers have already been operating in parallel and suddenly one of the delta groups develops a burn-out on one transformer leaving two good units for further operation. The first question to be asked is—what is the best thing to do? Or, what is the best combination to make to be in a position to take care of the biggest amount of energy put upon the whole of the remaining transformers? If it is kept in mind that it is impossible to get any more than 80 per cent. of the normal output per unit when an open-delta group is operating in parallel with a delta group of transformers, it will be an easy matter to know just how to proceed. Just what can be done and what ought to be done are given in Fig. 72 and Fig. 73. Supposing groups

No. 1 and No. 2 have been operating in closed delta and open-delta respectively and group No. 3 has just had a burn-out of one unit (see Fig. 72), it is quite evident that by connecting the three groups as shown more energy can be delivered than is possible with those connections shown in Fig. 73. In fact, it is possible to deliver more energy from six of the transformers shown in Fig. 72 than from the seven transformers with the connections given in Fig. 73.

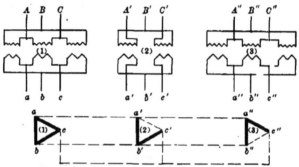

Fig. 73.—Wrong method of connecting open-delta transformers in parallel.

Single-phase vs. Three-phase Transformers.—As the art of transformer design and manufacture improves, the three-phase transformer is sure to be as extensively used as the single-phase transformer, especially so for high voltages; its only disadvantage being in the case of failure and interruption of service for repairs, but this will be offset by other important features since breakdowns will be of very rare occurrence.

From the standpoint of the operating engineer (neglecting the losses in the transformer) the single-phase transformer is at the present time preferable where only one group is installed and the expense of a spare unit would not be warranted as in the delta-delta system. If one of the three transformers should become damaged it can be cut out with a minimum amount of trouble and the other two can be operated at normal temperature on open-delta at approximately 58 per cent. of the total capacity of the three. With a three-phase transformer a damaged phase would cause considerable inconvenience for the reason that the whole transformer would have to be disconnected from the system before repairs of any kind could be made, which, in the case of a shell-type transformer, could probably be operated depending on the amount of damage, as it is not always possible to tell the exact extent of break-down before a thorough examination is made.

In the absence of any approved apparatus that can be relied on to take care of high-voltage line disturbances such as we have on some of our long-distance transmission lines, the whole burden being thrown on the insulation of this important link of a power undertaking, the three-phase transformer appears to be handicapped. Its break-down as mentioned above would entirely interrupt the service until a spare transformer is installed or the faulty one temporarily arranged with its faulty-short-circuited winding in the case of the shell-type. The engineer who has the responsibility of operating large power systems has not yet taken very favorably to the three-phase transformer for this very reason, his main object being reliability of service and not the first cost or saving of floor space.

It has for many years been appreciated by American and European engineers that apart from the decrease in manufacturing cost with increase in size of units, the three-phase unit has the advantage of requiring less material and is more efficient than any other single-phase combination of transformers of the same kw. capacity. The relative difference in the losses and weights being:

Three single-phase transformers weigh about 17 per cent. more than one three-phase.

Three single-phase transformers have about 17 per cent. more losses than one three-phase.

Used in open delta, two single-phase transformers weigh about the same as three single-phase transformers.

Two single-phase transformers have about the same losses as three single-phase transformers.

Used in tee, two single-phase transformers have a sum total weight of about 5 per cent. less than three single-phase transformers.

Two single-phase transformers have the sum total weight of about 5 per cent. less than two single-phase transformers connected in open-delta.

Two single-phase transformers have about 5 per cent. less losses than three single-phase transformers.

Two single-phase transformers have about 5 per cent. less losses than two single-phase transformers connected in open-delta.

Where a large number of transformers are installed in one building, say three groups or above, it is unquestionably a great saving over any combination of single-phase transformers, and the possibility of using two sets out of the three or three sets out of the four, and so on, offsets to a large extent the important drawback *reliability*, and places the three-phase transformer on almost an equal footing in this respect as the three single-phase transformer combination. The building is thereby reduced a considerable amount (also compartment insulator bushings and busbar high and low voltage insulators), besides simplifying the wiring layout in the stations.

To fulfil the requirements of a three-phase transformer using a combination of single-phase transformers, it is necessary to use:

Basis.—One three-phase transformer of 100 per cent. kw. capacity.

Delta Connected.—Three of 33.3 per cent. each, or a total 100 per cent. kw. capacity.

Star Connected.—Three of 33.3 per cent. each, or a total 100 per cent. kw. capacity.

Open-delta Connected.—Two of 57.7 per cent. each. or total 115.5 per cent. kw. capacity. (For three units, 173 per cent. kw. capacity is required.)*

"*Scott*": "*T*" (*Two-Transformer Connection*).—One of 57.7 per cent. kw. capacity, and one of 50 per cent. kw., or total 107.8 per cent. kw. capacity. (For three units, 165.6 per cent. kw.; capacity is required.)*

"*Taylor*": "*T*" (*Three-Transformer Connection*).—Three of maximum 50 per cent. each, or total 150 per cent. kw. capacity.

From this it is evident that the three best combinations are delta, star, and the three-transformer "T" connections. With the delta and "T" (three-transformer) systems a spare transformer is not warranted, and in case of a break-down of a unit, the minimum amount of time is lost in cutting it out of service. With the open-delta and "T" (two-transformer) systems, the loss of any unit stops the system from operating three-phase current. A further advantage of the three-transformer methods—delta, star, and "T"-three transformer—is, a spare unit costs less than one for either the open-delta or the "T"-two-transformer methods.

* Not always in service, hence a disadvantage.

THREE-PHASE TRANSFORMATION SYSTEM

CONNECTIONS FOR GROUNDING THREE-PHASE SYSTEMS

In Fig. 74 several methods are shown for grounding three-phase systems. For (A) and (B) there is a choice between the ground shown, or the ground at x. For (A) the ground as shown

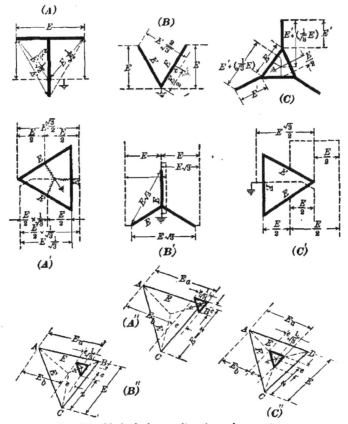

Fig. 74.—Method of grounding three-phase systems.

represents a maximum difference of potential between ground and line terminal of $\frac{1}{\sqrt{2}}E$, and $\frac{1}{\sqrt{3}}E$ for the ground at x.

For (B), the maximum voltage stress from line terminal to ground is full-line voltage, but with a ground made at x the maximum voltage stress is only approximately 87 per cent. of full voltage between terminals.

The delta-star system shown at (c) of Fig. 74 is the most uncommon of any of the systems given. Like (A'), three single-phase windings are required before grounding can be made possible. The maximum voltage strain from any line terminal to ground is $E' + (0.57\ E)$, or assuming $E = 100$, and $E' = 100$, we have $100 + (0.57 \times 100) = 157$ volts maximum strain from neutral ground to star-line terminal.

The diagrams of (A'), (B') and (C') of Fig. 74 represent the two well-known systems and their three methods of grounding. The two best methods for grounding are (A') and (B'). The method (C') is used only where auxiliary apparatus cannot be had for grounding through the star connection.

The delta and star systems shown at (B'') and (C'') represent systems grounded as shown at (A') and (B') but with one high-voltage terminal grounded. For (A'') the system is insulated with the exception of ground at high-voltage line terminal. Both B'' and C'' show the effects of grounding either the low voltage or high voltage or both. Where both neutral points are grounded, the high-voltage stresses on the low-voltage windings when an accidental ground occurs on the high-voltage side is reduced to a minimum, but is of maximum value on systems operating without grounded neutrals.

Three-phase to Single-phase Transformation.—An interesting three-phase to single-phase arrangement is given in Fig. 75. For this service three single-phase transformers or their equivalent (somewhat special in their construction) are required, the magnetizing current being much stronger than that used in the ordinary static transformer in order that the iron may be super-saturated. This modification will result not only in satisfactory transformation of voltage and current, but transformation of the frequency as well.

With three transformers connected as shown in Fig. 75, the secondary windings would under ordinary conditions show no e.m.f. across $A-C$; but if the iron is saturated the secondary becomes so transformed that an e.m.f. is obtained having 3f or three times the frequency of the primary.

The advantage of this arrangement is felt where arc and incandescent lighting is required and where the frequency is 15-25. The constant extension of electric traction on railways operating at these frequencies has resulted in the use of polyphase lamps and the triple frequency transformer arrangement. Single-phase railways are much more general than three-phase, and, therefore, it is really more important to be able to increase the frequency of single-phase than of three-phase current.

Also, in general, three-phase high voltages and energy are transmitted long distances to electric traction plants at a fre-

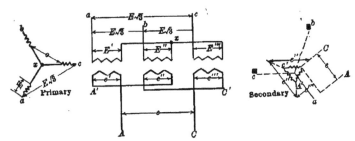

FIG. 75.—Method of transforming from three-phase to single-phase and changing the frequency.

quency of 60 cycles. It is therefore important from several viewpoints to be able to transmit electric energy at, say, $60 \div 3 = 20$ cycles and single-phase; thus reducing the voltage drop due to the higher frequency, and reducing cost of line construction, insulators and its maintenance in general.

CHAPTER V

THREE-PHASE TRANSFORMER DIFFICULTIES

Most of the troubles which occur on transmission systems are put down to line surges, resonance, or some unknown phenomenon on lines, and as a matter of fact most of the troubles might be in the transformers themselves, which may be damaged and their phase relations twisted so as to produce, in some instances, many times the normal voltage.

The most disastrous troubles that can happen to a three-phase system are those of complex grounds and short-circuits. With a grounded neutral star system, a ground on any one phase is a short-circuit of the transformers, and the entire group becomes disabled until changes are made.

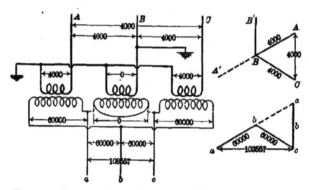

FIG. 76.—One transformer short-circuited and cut out of delta.

The voltage between windings and the core is limited to 57.7 per cent. of that of the line, and the insulation between the windings and the core is likewise reduced in proportion. The voltage between mains and the ground is 57.7 per cent. of the line voltage, with a star connection, but the neutral point may move so as to increase the voltage with an ungrounded system. If one circuit is grounded, the voltage between the other two circuits and the ground is increased, and may be as great as the

THREE-PHASE TRANSFORMER DIFFICULTIES 85

full line e.m.f. Such unbalancing would cause unequal heating of the transformers and if a four-wire three-phase system of distribution were employed, would prove disastrous to the regulation of the voltage.

With a star-delta system as shown in Fig. 76, where a transformer is short-circuited and cut out of delta on the secondary, it is possible to obtain $\sqrt{3}$ times the potential of any one of the transformers. In Fig. 76 *A B C* represents the vector triangle of e.m.fs. on the primary with full line voltage or 4000 volts, impressed on the transformers, which under normal conditions should be

$$4000 \times \frac{1}{\sqrt{3}} = 2300 \text{ volts.}$$

The phase relations are changed to 60 degrees, converting the original star arrangement to an open delta; one phase is

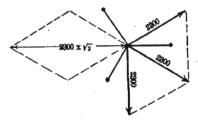

Fig. 77.—Primary e.m.fs. and phase relations.

reversed, the resultant e.m.f. being the same as that across any two phases. See also vector diagram, Fig. 77.

As each transformer is only designed for 2300 volts the e.m.f. across the secondary windings should be 34,600 volts, but in this case the voltages are 34,600 times $\sqrt{3}$ or 60,000 volts.

The secondary vector e.m.fs. are graphically represented to the right of Fig. 76. In order to bring the resultant vector secondary e.m.f., *a* and *c*, in its proper position the components must be drawn parallel with the primaries.

One secondary winding is short-circuited and cut out of delta leaving an open-delta connection reversed in direction, its phase relations being changed from 60 to 120 degrees; increasing the voltage between *a* and *c* to

$$34,600 \times 1732 \times 1732 = 103,557 \text{ volts.}$$
$$\text{or} \sqrt{3} \times \sqrt{3} \times 34,600 = 103,577 \text{ volts.}$$

This is a very important point to bear in mind, especially when generators are tied directly to the system without fuses or any protecting devices.

The voltages impressed on the primary windings of Fig. 76 and 77 are:

A and B = 4000 volts,
B and C = 4000 volts,
A and C = 4000 volts.

E.m.fs. between the primary neutral and any line, are:
A and A' = 4000 volts,
 which should be 6000×0.577 = 2300 volts;
B and B' = 0 volts,
 which should be 2300 volts;
C and C' = 4000 volts,
 which should be 4000×0.577 = 2300 volts.

The e.m.fs. between the secondary lines, are:
a and b = 60,000 volts,
 which should be $60,000 \times 0.577$ = 34,600 volts;
b and c = 103,577 volts,
 which should be $103,577 \div 2.99$ = 34,600 volts;
a and c = 60,000 volts,
 which should be $60,000 \times 0.577$ = 34,600 volts.

The increases in e.m.f. across the secondary lines, are:

a and c = 173 per cent. above normal,
b and c = 300 per cent. above normal,
a and b = 173 per cent. above normal.

It is also found that where the neutral points of the primary and secondary windings are grounded, the opening of one or two of the three line circuits will cause currents to flow through the ground. A partial ground on a line circuit will partially short-circuit one transformer and cause current to flow through the ground and the neutral.

The actual strain between high-tension and low-tension windings is equal to the high-tension voltage plus or minus the low-tension voltage, depending upon the arrangement and connection of the coils; but as the low-tension voltage is usually a small percentage of that of the high-tension, it is customary to assume that the strain between windings is equal to that of the high-tension voltage alone.

If the neutral points of the high-tension and low-tension windings are grounded, the iron core being also grounded, then

THREE-PHASE TRANSFORMER DIFFICULTIES 87

as long as the circuits are balanced the voltage strains will be the same as with the windings ungrounded, and balanced; but in case of a ground on either high-tension or low-tension line, or in case of a connection between high-tension and low-tension windings, a portion of the windings will be short-circuited.

Assuming that all lines and transformers are in good shape, that is to say, clear from grounds and short-circuits, it is possible to obtain any of the following results shown in Figs. 78, 79, 80 and 81, by connecting the receiving ends of transmission lines to a wrong phase terminal receiving three-phase current from another source of supply, or by switching together groups of two or more transformers of the wrong phase relations.

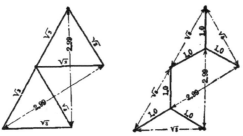

FIG. 78.—Resultant e.m.fs., and phase relations of improper delta-delta and star-star connected group of transformers.

Fig. 78 represents the result of a delta-delta and star-star combination thrown together at 120 degrees apart, both transmission lines receiving three-phase currents of the same potential, phase relations, and frequency.

The resultant voltage obtained in attempting to parallel two groups of three transformers star connected is $\sqrt{3}$ times the e.m.f. between any two line wires, or

Star $= (57.7 \times 1.732 = 100) \times (1.732) = 173.2$ volts.

The combination shown in Fig. 79 represents four groups (three single-phase transformers in each group) connected to one set of busbars. Each group receives three-phase current from independent source of supply and is so tied in on the primary and secondary busbars as to involve a partial short-circuit.

In common practice this combination is more often likely to happen on large distributing systems where all transformers in

88 STATIONARY TRANSFORMERS

groups are tied together on primaries and secondaries. As will be noticed, any attempt to connect such a system with all primary windings and all secondary windings of each group in parallel will produce a short-circuit.

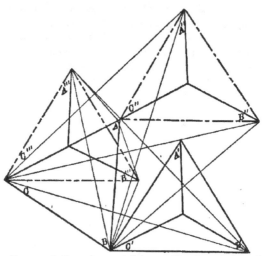

Fig. 79.—Representation of a complete combination of delta-delta and delta-star transformer group connections.

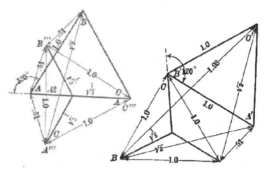

Figs. 80 A and B.—Graphic illustration of e.m.fs. and phase displacement of two delta-delta to delta-star connected groups of transformers.

With a delta-star presupposed parallel operation it is impossible to change the magnetic field to correct the phase displacement

THREE-PHASE TRANSFORMER DIFFICULTIES 89

which occurs, though it is possible in the case of generators which are necessary for permitting the 30 degrees electrical displacement to be corrected by a mechanical twisting of the phases with respect to their magnetic fields; but with transformers it is impossible.

The phase displacements show a star connection introduced in which the relative e.m.f. positions are changed by an angle 30 degrees. If, for example, we assume the line potential to be 60,000 volts, and we attempt to connect the groups as shown in diagram, the result will be voltages as high as 116,000.

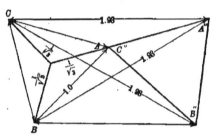

FIG. 81.—E.m.fs. and phase relation of a delta-delta to delta-star connected group of transformers.

The resultant e.m.fs. established by this experiment, are shown separately in Figs. 80 and 81. They are correctly:

B'' to B = 84,840 volts, which should be = 0
B'' to C = 116,000 volts, which should be = 60,000
A'' to C = 116,000 volts, which should be = 60,000
A'' to B = 116,000 volts, which should be = 60,000
A' to B = 84,840 volts, which should be = 60,000
A' to C = 31,000 volts, which should be = 60,000
C' to B = 116,000 volts, which should be = 60,000
C' to B = 116,000 volts, which should be = 60,000
C' to C = 84,840 volts, which should be = 60,000
B''' to B = 31,000 volts, which should be = 0
B''' to A = 31,000 volts, which should be = 60,000
A''' to A = 31,000 volts, which should be = 0
A''' to B = 84,840 volts, which should be = 60,000

A fact not very well recognized, is the impossibility of paralleling certain primary and secondary three-phase systems. The following combinations can be operated in parallel:

Delta-star group with a delta-star.
Star-star group with a star-star.
Delta-delta group with a delta-delta.
Star-delta group with a star-delta.
Delta-star group with a star-delta.
Star-star group with a delta-delta.

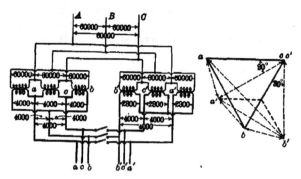

FIG. 82.—Practical representation of a delta-delta to delta-star connected group of transformers.

These delta-star combinations necessitate changing transformer ratios of primary and secondary turns, as:

Delta-star is a constant of $\sqrt{3} = 1.733$ to 1.

Star-delta is a constant of $\dfrac{1}{\sqrt{3}} = 0.577$ to 1.

consequently, special ratios of secondary to primary turns are needed in delta-star or star-delta transformers, in order to produce standard transformation ratios.

As already explained, displacement of phase relations occur on the secondary side of transformers when two or more groups are connected delta-delta and a delta-star; the delta-delta having a straight ratio and the delta-star a ratio of 1 to 0.577; or any of the following combinations:

Delta-star ratio $\dfrac{1}{\sqrt{3}}$ to 1, and star-star with ratio 1 to 1, or any ratio.

Delta-delta, ratio 1 to 1, with a star-star ratio $\dfrac{1}{\sqrt{3}}$ to 1, or any ratio.

Star-delta, ratio 1 to $\frac{1}{\sqrt{3}}$, and star-star with ratio 1 to 1, or any ratio.

The phase relations occupy a relative shifting position of 30 degrees on one group of transformers to that of the other. A more practical representation of the secondary voltages and phase relations of a delta-delta and delta-star is shown in Fig. 82.

In order to tie in a large number of local existing plants consisting of gas, steam and water-driven generators, etc., with two- and three-phase distributing systems, special care and thought are required in laying out the right scheme of connections. One sometimes meets with a two-phase three- and four-wire distribution and several other kinds of systems of odd voltages and frequencies which must be tied in on the main high-voltage transmission system through transformers (and probably frequency changes), all of which requires special knowledge on the part of those whose duty it is to operate them.

Consolidated systems of this kind generally have to contend with parallel operation of local power plants, this sometimes being done directly from the high voltage side or line side of the power transformers for voltages as high as 60,000 volts, by the application of potential transformers. Only a small number use this method of synchronizing their auxiliary stations and it is not considered good practice.

Before parallel operation of any kind is done it is always advantageous to know all about the connections and voltages of the different transformer groups. With the delta and star systems it is (as has already been explained) only possible to parallel six combinations out of the ten so commonly used. Of these six combinations the transformers to be paralleled must have equal impedance and equal ratio of resistance to impedance. With equal impedance the current in each unit will be in proportion to their rated capacity in kw.'s, although the sum of the currents may be greater than the current in the line; if, on the other hand, the impedance of the units is unequal, the current in each unit will be inversely proportional to its impedance; that is to say, if one unit has 1 per cent. impedance and the other 2 per cent. impedance, the first unit will take twice as large a per cent. of its rated capacity as the second unit—the sum of the currents in the two units may be or may not be equal to the line current. With equal ratios of resistance to reactance the current in each unit will be in phase with the current in the line, also the sum of the currents will be the same as the line current. With unequal ratios of resistance to reactance the current in each unit will not be in phase with the current in the line, therefore the sum of the currents will be greater than the line current. If, however, the impedance of the units is the same, both will carry the same per cent. of full-load current; and if, in addition, the ratio of resistance to reactance is the same in both cases, the current in the two units will be in phase with each other, and their numerical sum will equal the load current, thus *there will exist perfect parallel operation.*

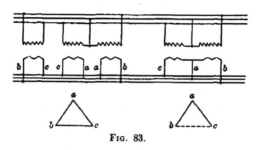

Fig. 83.

Whatever scheme of transformers is decided on for a given system it will always be advisable to keep to that scheme if possible; this is particularly applicable on some of the larger systems of 100 megawatts and over where networks of high-voltage transmission lines and sub-stations are numerous.

Quite a number of systems have distance sub-stations with only two groups of transformers, both groups being operated in parallel at all times. Assuming the transformers to be connected in delta-delta and one transformer of one of the groups becomes damaged, it might mean, if the load is great, that at least one of the other group must be cut out. If the load is not greater than 80 per cent. of their total rating it will be possible under ordinary circumstances to operate the closed delta group with the open-delta as shown in Fig. 83.

It is not generally known that there are sixteen different connections on one transformer star-connected group, and that these connections can be changed about to obtain several parallel combinations, such as those shown in Fig. 84 giving the time-phase of 0, 30, 60, and 90 degrees (electrical).

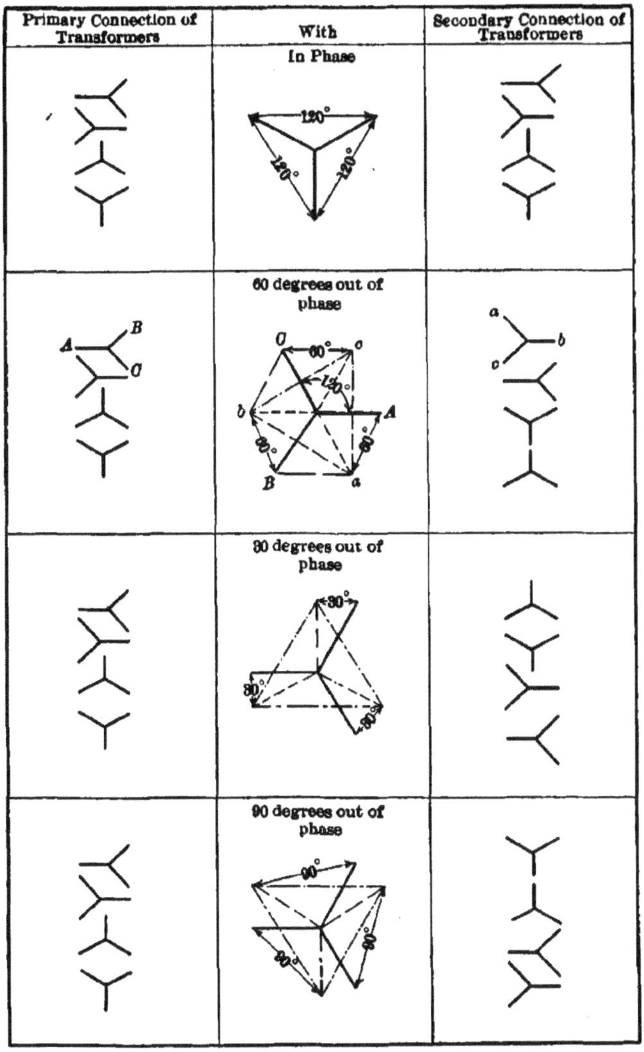
Primary Connection of Transformers
With
Secondary Connection of Transformers
In Phase
120°
120°
120°
60 degrees out of phase
A
B
C
C
60°
c
120°
b
A
60°
60°
B
a
a
b
c
30 degrees out of phase
30°
30°
30°
90 degrees out of phase
90°
90°
90°

Fig. 84.

94 STATIONARY TRANSFORMERS

Not including the straight star or delta connection with a time-phase angle equal to zero, we find that there are no less than three different groups such that transformers belonging to the same group can be connected in parallel, while if of different groups, no parallel connection is possible without a special rearrangement of the internal connections of the transformers.

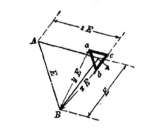

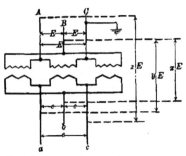

Fig. 85.—High-voltage line grounded, producing maximum strain on low voltage windings. (Neutrals non-grounded.)

The troubles usually experienced with high-voltage transformers can be classified as:

(a) Puncturing of the insulation between adjacent turns due to surges, etc.

(b) Shifting of coils due to switching on and off very heavy loads.

(c) Terminals puncture (transformer insulator bushing and other leads), due to either (a) or (b) or both.

There are a large number of causes for transformer breakdowns, some of them being:

(1) Insufficient insulation between layers and turns.

THREE-PHASE TRANSFORMER DIFFICULTIES

(2) Insufficient insulation on the end-turns.

(3) Electromagnetic stresses too great.

(4) Electrostatic capacity of certain parts too high.

(5) Condenser effect between coils, and between windings and ground too high.

(6) Improper drying out after construction.

(7) For want of a choke coil or reactance in series.

(8) Concentrated condensers in parallel with transformer windings.

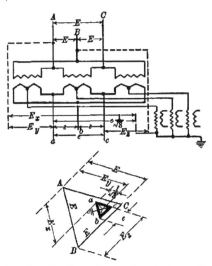

Fig. 86.—Showing the effect of grounding the low-voltage neutral.

(9) Two or more of the above in connection with concentrated condensers.

(10) Internal short-circuits.

(11) Conditions of switching, surges, arcing grounds, lightning, etc.

(12) Improper treatment of the oil.

(13) Oil not suited to the transformer.

(14) Thickening of oil and clogging of cooling medium stopped.

(15) Leaking water-coils, or, circulation of cooling medium stopped.

(16) Breathing action (prevalent in damp localities).

STATIONARY TRANSFORMERS

(17) Improperly installed protective apparatus.
(18) High winds which bring live conductors in oscillation.
(19) Defective governors or prime-movers.
(20) Variation in speed of generators.
(21) Variation in generator voltage.
(22) Roasting by constant over-load.
(23) Puncturing of transformer terminals.
(24) When transformers are connected to generating stations and systems and bus-bars having a total kw. capacity many times greater.

Figs. 85, 86, and 87 show other causes for break-downs in high-voltage transformers. Three different conditions of operation are given; one for an insulated delta-delta system showing the effects of a grounded live conductor, one for a delta-delta system with the neutral point of low-voltage windings grounded, and the other for a delta-delta system with both high- and low-voltage windings grounded. The approximate maximum potential strains are shown by the vector relations. These serve to show other greater difficulties which receiving-station transformers are subjected to in addition to those usually impressed on the transformers at the generating stations.

CHAPTER VI

THREE-PHASE TWO-PHASE SYSTEMS AND TRANSFORMATION

With two or three single-phase transformers it is possible to have three-phase primaries with two-phase secondaries, or *vice versa*. For long-distance transmission of electric power the three-phase system is universally adopted because it requires less copper for the line than either the single-phase or the two-

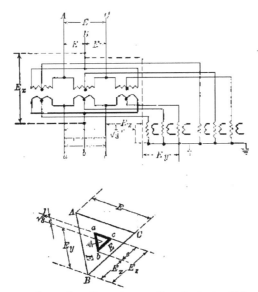

Fig. 87.—This shows the effect of grounding both the high- and low-voltage neutrals, resulting in minimum voltage strain.

phase systems to transmit a given amount of power with a given line loss, and with a given line voltage. The two-phase system offers certain advantages over the three-phase system when applied to local distribution of electric power.

In Fig. 88 is shown the well known three-phase three-wire to two-phase four-wire transformation. Two transformers are all that is necessary in this arrangement, one of which has a 10 to 1 ratio and the other a 10 to 0.866, or 10 to $\frac{\sqrt{3}}{2}$.

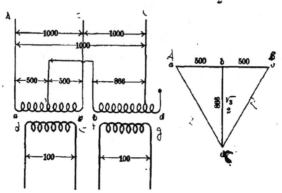

Fig. 88.—The three-phase two-phase connection* (Scott system).

One wire, b, of the 10 to 0.866-ratio transformer is connected to the middle point of the 10 to 1 ratio transformer, the ends of which are connected to two of the three-phase mains, $a\ c;\ d$, the

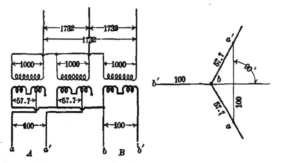

Fig. 89.—Three-phase to two-phase star-connected transformers.

end of the other transformer is connected to the remaining wire of the three-phase mains.

It is customary to employ standard transformers for the three-phase two-phase transformation, the main transformer having a ratio of 10 to 1, and the other transformer a ratio of 9 to 1.

* Patent No. 521051, June 5th, 1894.

THREE-PHASE TWO-PHASE SYSTEMS

By a combination of two transformers it is possible to change one polyphase system into any other polyphase system.

The transformation from a three-phase to a two-phase system may be effected by proportioning the windings, as shown in Fig. 89. The three transformers are wound with a ratio of transformation of 10 to 1. The secondaries of two of the transformers have two taps each, giving 57.7 per cent. and full voltage, so that they serve as one phase of the two-phase transformation. The primary windings are connected in star.

The secondary windings are also connected in star. In Fig. 89, $b\,b'$ represents the secondary voltage from b to b' in one trans-

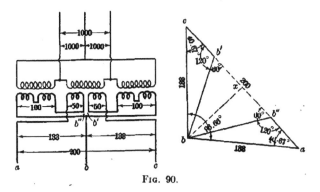

Fig. 90.

former. At an angle of 90 degrees to $b\,b'$ the line, $a\,a'$, represents in direction and magnitude the voltage, a to a', which is the resultant of the two remaining transformer e.m.fs., giving 57.7 per cent. of the full voltage, $57.7 \times \sqrt{3} = 100$. From the properties of the angles it follows that, at the terminals, $a\,a'$ and $b\,b'$, two equal voltages will exist, each differing from the other by 90 degrees, and giving rise to a two-phase current.

Still another method of getting two-phase from three-phase, or *vice versa*, consists in cutting one phase, say (Fig. 90) the middle transformer of the delta-connected group, in half, and arranging one-half to the left at $b'\,c$, and the other half to the right at $b''\,a$.

The resultant of $a\,b'$ and $b'\,b$ is one side, or one phase of the two-phase transformation.

The resultant of $a\,b''$ and $b''c$, is the other phase of the two-phase transformation.

It is evident as shown in Fig. 90 that the two-phase relation is a trifle over 90 degrees; since the angle, $b\ b''\ x$ and $b\ b'\ x$, is 60 degrees, and the sine of 60 degrees is equal to $\frac{\sqrt{3}}{2} = 0.866$, the tangent of the angles, $b\ c\ x$ and $b\ a\ x$, are likewise $\frac{\sqrt{3}}{2} = 0.866$. Therefore, the angle, $a\ b\ c$, must equal 90 degrees nearly.

The angle, $b\ c\ a$, whose tangent is 0.866 is an angle of 40.67 degrees; therefore,

$(b\ c\ x = 40.67) + (b\ a\ x = 40.67) + (a\ b\ c = 98.66) = 40.67 + 40.67 + 98.66 = 180$ degrees, nearly.

The approximate voltage obtained between $c\ b$ and $b\ a$ is 133 volts.

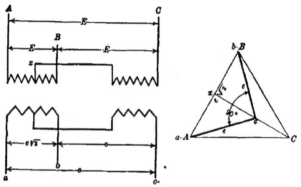

FIG. 91.—Three-phase to three-phase two-phase* (Steinmetz system).

With two or more transformers it is possible to transform from three-phase to two distinct phase currents of three-phase and two-phase systems. In the arrangement shown in Fig. 91 only two transformers are used. The two primary windings are connected to the three-phase mains. One transformer is wound with a ratio of transformation of 10 to 1. The other with a ratio of 0.866 to 1. The primary and secondary windings of this transformer are connected to the middle of the primary and secondary windings, respectively, of the first.

$a\ b$ represents the secondary voltage from a to b in one transformer. At right angles to $a\ b$ the line, $x\ c'$, represents in direction and quantity, the voltage, x to c', of the second transformer.

* Patent No. 809996, January 16th, 1906.

At the terminals, $a\,b\,c$, three equal voltages will exist, each differing from the other by 60 degrees, and giving rise to a three-phase current.

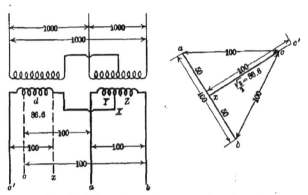

FIG. 92.—Three-phase "T" to two-phase four-wire.

It also follows that, at the terminals $a\,b$ and $x\,c'$, two equal voltages will exist, each differing from the other by 90 degrees, and giving rise to a two-phase current. As will be noted, the

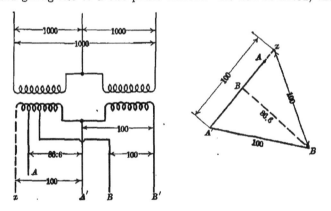

FIG. 93.—Three-phase open delta to three-phase two-phase (Taylor system).

voltages obtained in the three-phase side are equal to those between any phase of the two-phase system.

The arrangement in general is similar to that of the ordinary "V" or open-delta system.

Another combination somewhat similar to the above is shown in Fig. 93. The primary windings of the two transformers are connected in open-delta. The secondary windings are connected in such a manner as to give two distinct phase currents; one kind differing in phase by 90 degrees, and the other by 120 degrees. From one secondary winding two special taps of 50 per cent. and 86.6 per cent. are brought out to complete the circuits of three-phase and two-phase secondary. By this method of connection it is possible to obtain two-phase currents from $A\ A'$ and $B\ B'$, also three-phase currents from $x\ A'$, $A'\ B$, and $B\ x$, the two-phase e.m.fs. will be 86.6 per cent. of those of the three phase.

The method shown in Fig. 94 is a device patented by the writer,

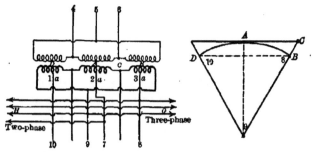

FIG. 94.—Three-phase delta to three-phase two-phase* (Taylor system).

and employed to operate both two-phase and three-phase electric translating devices, on one four-wire system of distribution; and to operate independent systems in parallel circuit on said four-wire system.

Three single-phase transformers are used. The primary windings are shown connected in delta, and the secondary windings also connected in delta. A distribution line, 7, tapped at the middle of the secondary winding, $2a$; a distribution line, 8, tapped at $\dfrac{\sqrt{3}}{2}$ per cent. of the length from one end of the winding, $3a$; a distribution line, 9, tapped on the connection between the end windings, $2a$ and $3a$; a distribution line, 10, tapped at $\dfrac{\sqrt{3}}{2}$ per

* Patent No. 869595, October 29th, 1907.

THREE-PHASE TWO-PHASE SYSTEMS

cent. of the length from the end of winding 1a, and translating means connected on said distribution lines both for two- and three-phase on.

At 1a, 2a and 3a are the secondary windings of said transformers. The secondary winding, 1a, is tapped at D, which is about 86.6 per cent. of its length, by the line, 10; which serves also as a leg for both the two- and three-phase circuits.

The secondary winding, 2a, is tapped at its central point A, by a line, 7; forming one leg of the two-phase circuit. The secondary winding, 3a, is tapped at approximately 86.6 per cent. of its length from one end, at about the point, B, by a line, 8, to serve

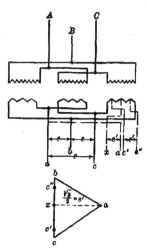

FIG. 95.—Three-phase to two-phase, giving 86 per cent. standard transformer taps.

as one leg of both the three- and two-phase circuits. C represents the point of a tap taken from the junction of two secondary windings which are shown connected in the series circuit, which serves as another leg for both two- and three-phase circuits.

The arrangement accomplishes the operating of non-synchronous apparatus of two-phase and three-phase design without the aid of transformers or split-phase devices. The operation consists in generating three-phase alternating currents in the lines, 4, 5 and 6, transforming the same into three-phase currents in the legs, 8, 9 and 10, and into two-phase currents in the legs, 7, 8, 9 and 10; and in operating translating devices

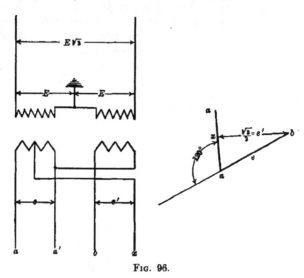

Fig. 96.

at G and H in parallel with the two- and three-phase current circuits. The two-phase windings used on the motors must be

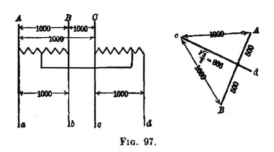

Fig. 97.

independent as the interconnected type of winding would not operate on this system.

A three-phase single-phase transformation is shown in Fig. 98.

The objection to this connection is the distorted effect of the relative voltages and phase relations of the three-phase when a single-phase load is put on one of the phases. To obviate this to some extent it would be necessary to give the three-phase voltage a slight distortion. The unbalanced voltages and phase relation when a single-phase load is applied is shown by the vector in Fig. 98, it having the effect of twisting the phase relation when a load is applied between b–e from a symmetrical point as shown at the point a'.

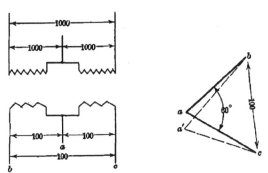

FIG. 98.—Three-phase to single-phase.

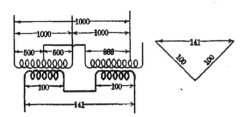

FIG. 99.—Another three-phase single-phase secondary operation.

It is possible to take currents from a three-phase system and transform them into a single-phase current (see Fig. 99). All that is necessary is to arrange two transformers so that their connections are identical with the ordinary two-phase to three-phase transformation, the only difference being in the secondary, which has the two windings connected in series for supplying a single-phase circuit.

In the ordinary three-phase to two-phase transformation, the two components in each half of the winding differ in phase by 90

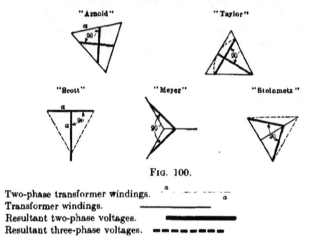

Fig. 100.

Two-phase transformer windings.
Transformer windings.
Resultant two-phase voltages.
Resultant three-phase voltages.

degrees. However, when the secondary circuits are connected in series, these two component currents are of one phase.

The majority of three-phase two-phase transformer connections employed are for temporary or special purposes. This is particu-

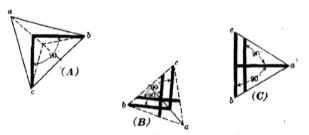

Fig. 101.—Combination of systems given in Fig. 100.

larly so when only two transformers are used. The systems used in Europe and America are shown in Fig. 100, in the form of vectors. In comparing these various systems by means of the vectors given, it is very interesting to note how near they are to being one and the same thing. The relations are about the same but the transformers in each case are connected quite differently.

The combinations are shown in Fig. 101. In (A) the Meyer and Steinmetz are combined into one; in (B) the Arnold and Steinmetz are combined, and in (C) the Scott and Taylor systems are shown combined into one. All of the systems given here serve to show their likeness and are interesting to all those who might be in need of emergency substitutes.

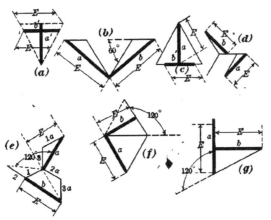

FIGS. 102 and 103.—Other three-phase two-phase methods using standard transformer.

Besides these systems given in Figs. 100 and 101 there are other connections of less value but nevertheless important in that they might be found useful in some particular instant of break-down of existing apparatus, or for temporary purposes. These are shown in Figs. 102 and 103, where (a) is composed of two single-phase transformers of 10 to 1 and 9 to 1 ratio respectively, or of two 10 to 1 ratio transformers, one of the transformers being tapped at the 9 to 1 ratio point and connected to the 50 per cent. point of the other transformers. (b) is the ordinary delta connection with the two halves of one winding or one single-phase unit reversed. (c) is the ordinary delta system with taps taken off as shown. (d) is the ordinary star with one transformer secondary winding cut into three equal parts. (e) is a "distributed" star, 1–1a, 2–2a and 3–3a representing three

108   STATIONARY TRANSFORMERS

single-phase transformers with winding cut into two equal parts and connected as shown. (*f*) is a star connection with one transformer winding reversed. (*g*) is an ordinary open-delta connection. With the exception of (*a*) and (*g*) which consist of two single-phase transformers, all the others are three-phase two-phase combinations using three single-phase transformers.

By the use of transformers other than standard ratios and design, the three-phase two-phase transformer combinations shown in Fig. 104 can be made.

CHAPTER VII

SIX-PHASE TRANSFORMATION AND OPERATION

In transforming from three-phase to six-phase there are four different ways of connecting the secondaries of the transformers: namely, diametrical—with or without the fixed neutral point; double star; double delta; and double tee. In the first three cases the primaries may be connected either star or delta, according to the voltage that each winding will stand, or to obtain a required

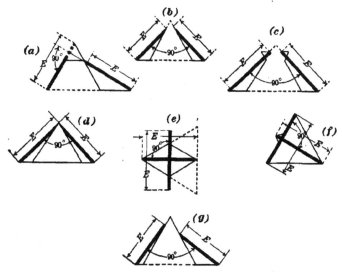

Fig. 104.—Three-phase to two-phase using special transformer tops.

secondary voltage. In the last case, the primary windings are connected in tee.

For the diametrical connection three single-phase transformers may be used with one central tap from each transformer secondary winding, or there may be six secondary coils. For the double-star or double-delta connection two independent secondary coils are required for each transformer; the second set are all reversed, then connected in a similar manner to the first set. Hence, the phase displacement is shifted 180 degrees.

For the double-tee connection two single-phase transformers are required, one of which has a 10 to 1 ratio and the other a 10 to 0.866, or 10 to $\dfrac{\sqrt{3}}{2}$ ratio.

There are two secondary coils giving 10 to 1 ratios, and two giving 10 to 0.866 ratios.

In six-phase circuits there are coils with phase displacements of 60 degrees; each coil must move through 180 electrical degrees from the position where the current begins in one direction, before

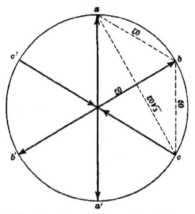

FIG. 105.—Six-phase diametrical e.m.fs. and phase relation.

the current begins to reverse. Hence, for the double star, diametrical, double delta and double-tee connections if the ends of the transformer coils are reversed, the phase displacement of the e.m.f. is in effect shifted 180 electrical degrees.

Take, for instance, the e.m.fs., $a\,a'$, $b\,b'$ and $c\,c'$, as graphically explained in Figs. 106 and 107, for a diametrical connection they are equal to $2\,a\,x$, $2\,b\,x$, $2\,c\,x$, etc.

For double-star connection:

$a\,b'$, $b'\,c$, $c\,a$, etc., is $\sqrt{3}$ times $x\,a$, $x\,b'$, $x\,c$, etc.

For double-delta connection:

$x\,a$, $x\,b'$, $x\,c$, etc., is $\dfrac{1}{\sqrt{3}}=0.577$ per cent. of $a\,b'$, $b'\,c$, $c\,a$, etc.

For double-tee connection:

SIX-PHASE TRANSFORMATION AND OPERATION 111

$a\ b'$, $b'\ c$; $c\ a$, etc., is 13.3 per cent. more than $a'\ y$, or $a\ z$.

The general statement of relationship between e.m.fs. in Fig. 106 may demonstrate that if the value of $a\ a'$, etc., is represented by the diameter of a circle, the values of $a'\ b$, etc., are repre-

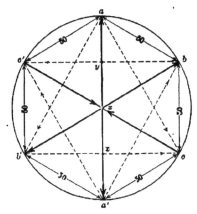

FIG. 106.—Six-phase e.m.fs. graphically represented.

sented by a 120-degree chord, and the values of $a\ b$, etc., are represented by a 60-degree chord of the same circle.

If the voltage between, $a\ b$, etc., is not required, only three secondary coils are needed; but if this voltage should be required,

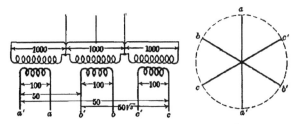

FIG. 107.—Six-phase diametrical connection.

then six secondary coils are needed, or three coils with a center tap like that shown in Fig. 107. The diametrical connection of transformer secondaries as represented in Fig. 107 is the most commonly used of any three-phase to six-phase transformations. One secondary coil on each step-down transformer is all that is

112 STATIONARY TRANSFORMERS

necessary; whereas the double-star, double-delta, and double-tee connections require two secondary coils, and therefore four secondary wires for each transformer.

The two secondary wires from each transformer are connected to the armature winding of a rotary converter at points 180 degrees apart—such as shown at $a\ a'$, $b\ b'$, $c\ c'$; therefore, arrangements for the diametrical connection are much simpler than any of the others.

A part of the three-coil secondary diametrical connection may be used for induction-motor service to start the rotary converter,

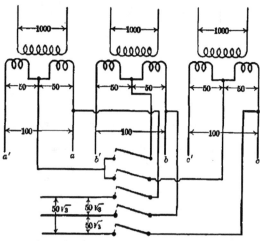

Fig. 108.—Six-phase diametrical connection with five-point switch used in connection with motor for starting synchronous converters.

and when sufficient speed is obtained the motor may be cut out of service. The arrangement is shown in Fig. 108. By means of this connection, which is made through the introduction of a five-pole switch, a three-phase e.m.f. may be obtained, giving a value equal to half the e.m.f. of each secondary winding times $\sqrt{3}$. That is to say, if half the e.m.f. of each secondary winding is equal to 50 volts, then assuming the switch to be closed, we obtain $50 \times \sqrt{3} = 86.6$ volts.

Similar ends of the three windings are connected to three points on one side of the five-pole switch. The three wires on the other

SIX-PHASE TRANSFORMATION AND OPERATION

side of the switch are led off to the three-phase motor service. The two remaining points of the switch receive three wires from the neutral points of the three secondary windings. Connections are so made that when the switch is closed a star-connection is obtained.

With the double-star arrangement of secondary windings, shown in Fig. 109, a rotary converter may be connected to a given three of the six secondary coils, or one rotary may be connected to the six secondary coils. The disadvantage of star connection is that in case one transformer is burned out, it is not possible to continue running.

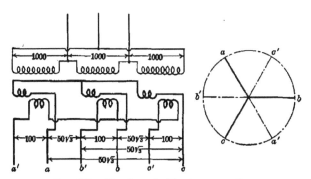

Fig. 109.—Six-phase double-star connection.

An arrangement for six-phase transformation is shown in Fig. 110, which differs from that of Fig. 109 in that the middle point of each transformer winding is tied together to form a neutral point for the double star combination.

It is common practice to connect the neutral wire of the three-wire, direct-current system to the neutral point of the star connection.

It may be seen that the similar ends of the two coils of the same transformer or similar ends of any two coils bearing the same relation to a certain primary coil are at any instant of the same polarity.

The double-delta secondary arrangement should preferably be connected delta on the primary, as it permits the system to be operated with only two transformers, in case one should be cut out of circuit.

One set of the three secondary coils is connected in delta in the ordinary way, but the leads from the second set are reversed and then connected in a similar manner.

It can be seen from Fig. 111 that two distinct delta connections are made, and in case it is desired to connect the six leads, $a\ b\ c$-$a'\ b'\ c'$, to a six-phase rotary converter it is necessary that each be connected to the proper rings.

The double-tee connection requires only two transformers, and so far as concerns the cost of the equipment and the efficiency in

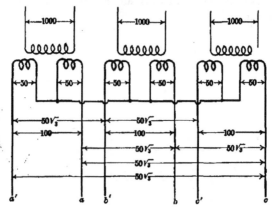

Fig. 110.—Six-phase double-star with one neutral point for the six secondary windings.

operation two tee-connected transformers are preferable to the delta or star connections. This connection can be used to transform two-phase to six-phase, and from three-phase to six-phase.

It is worthy of note that the transformer with the 86.6 per cent. winding need not necessarily be designed for exactly 86.6 per cent. of the e.m.f. of the other transformer; the normal voltage of one can be 90 per cent. of the other, without producing detrimental results.

Fig. 112 represents the tee-connection for transforming from three-phase to six-phase e.m.fs.

With reference to its ability to transform six-phase e.m.fs. and maintain balanced phase relations, the tee-connection is much better than either the delta or star connections.

Another interesting method of transforming from three-phase or two-phase to six-phase is shown in Fig. 113. The two trans-

SIX-PHASE TRANSFORMATION AND OPERATION 115

formers bear the ratio of 10 to 1 and 10 to 0.866, as explained in the previous example. For a two-phase primary supply, $A\ A'$ is tapped on one phase, and $B\ B'$ is tapped on the other; the line, x, being cut loose. For a three-phase primary supply the lines,

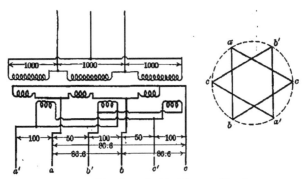

Fig. 111.—Six-phase double-delta combination.

A, A', and B, are connected to the three-phase mains. The secondary connections in both cases remain the same.

If a neutral wire is required as in the case of the three-wire, direct-current system, it may be taken from the point, y. For

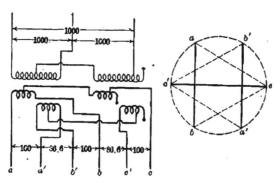

Fig. 112.—Six-phase tee connections.

running a blower motor, or to furnish current for running the rotary converter up to synchronous speed by an induction motor mounted on the same shaft, any one of the two secondary tee-connections may be used. The three-phase e.m.f. obtained would have a value equal to the full secondary voltage used for the rotary converter.

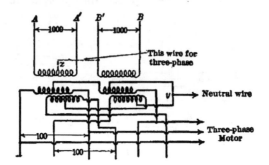

Fig. 113.—Six-phase from three-phase or two-phase.

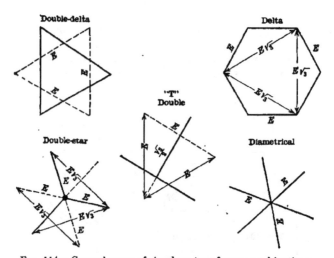

Fig. 114.—General group of six-phase transformer combinations.

Fig. 114 shows all the six-phase transformer combinations in use at the present time.

CHAPTER VIII

METHODS OF COOLING TRANSFORMERS

Small transformers do not require special cooling devices since they have large radiating surface compared with their losses. Large transformers will not keep cool by natural radiation; some special cooling devices must be provided.

The various cooling methods are:
Self-cooling dry transformers.
Self-cooling oil-filled transformers.
Transformers cooled by forced current of air.
Transformers cooled by forced current of water.
Transformers cooled by forced current of oil.
Transformers cooled by some combination of above means.

Self-cooling Dry Transformer.—Transformers of this kind are usually of small output, and do not require any special means of cooling, the natural radiation being depended upon for cooling.

Self-cooling Oil Transformers.—This arrangement is employed for at least 60 per cent. of transformers in use, the core and coils being immersed in oil. The two advantages gained by immersing these transformers in oil, are: Insulation punctures can in many cases be immediately repaired by the inflow of oil, and the temperature is reduced by offering means of escape for the heat.

Many manufacturers depend upon the high insulating qualities of the oil itself, and, therefore, introduce less insulating material such as cambric and mica, etc. On the other hand if oil is punctured it will close in again, unless the puncture be the result of a short-circuit in the transformer, in which case an explosion is liable to occur, or a fire started. In this way electric plants have been destroyed.

For the purpose of obtaining the necessary radiating surface, tanks of the large self-cooled transformers of many manufacturers are made of thin corrugated steel or cast iron. The thin corrugated steel metal tanks are not sufficiently strong to be safely handled in transportation with the transformers in them. A slight blow is sufficient to cause the oil to leak at the soldered seams between the different sheets of thin steel, or at the joints between the sides. The cast-iron case is unquestionably the best, and the most suitable for oil transformers; the great strength and stability of cast-iron cases insure the safe transportation of the transformer.

In the design of oil-insulated transformers, interior ventilation is provided by oil passages or ventilating ducts, between the coils, and in the iron. These secure an even distribution of heat and a uniformity of temperature throughout the transformer, results which can be secured only by a free internal circulation of oil.

Without good oil circulation, transformers of large size may reach an internal temperature greatly in excess of that of the external surface in contact with the oil, and in poorly designed transformers this may lead to the speedy destruction of the insulation of the coils.

The number and size of the oil passages, or ventilating ducts are planned to keep all parts of the transformer about evenly cooled. Such ducts necessarily use much available space and make a transformer of a given efficiency more expensive than if the space could be completely filled with copper or iron. Experience with oil-insulated transformers of large size and high voltage has shown that oil increases the life of the insulation, in addition to acting as a cooling medium, and adds materially to the capability of the transformer to resist lightning discharges, in other words such a transformer is safer than a dry transformer.

The amount of heat developed in a transformer depends upon its load and its efficiency. In a 500-kw. transformer of 98.5 per cent. efficiency there is a loss at full load of 7.0 kw. Since this loss appears as heat, it must be disposed of in some way or the temperature will rise until it becomes dangerously high.

The self-cooled oil-insulated transformer is now made in sizes up to 3000 kv-a capacity and represents one of the best advances in the manufacture of transformers. This new design is a very important development of the art of transformer manufacture. This type will be more in demand than the air-cooled type (air-blast) where water is or is not available, is expensive or not sufficient. It requires a minimum amount of attention and no auxiliary apparatus or equipment. The great problem to be solved in this type was that of providing sufficient surface to radiate the heat generated and keep the temperature rise within

METHODS OF COOLING TRANSFORMERS 119

certain limits. The total amount of surface broken up into corrugations depends on its efficiency which may be defined as the *watts radiated per square inch of surface*, as well as on the amount of heat to be radiated. A plain surface is found to be the most efficient as both the air and oil come into close contact with it, but as it is broken up into corrugations, the efficiency is decreased slightly. The most modern self-cooled type is provided with auxiliary pipes or radiators whereby the actual surface of the tank can be greatly increased and at the same time the radiating efficiency of the surface kept very high. The method of cooling consists essentially of fitting the outside of a plain cast-iron or plain boiler-plate tank with a number of vertically arranged tubes, the upper ends of which enter the tank near the top and the lower ends near the bottom.

An idea of the limitations in this direction can be best obtained by making a rough comparison of two transformers of widely different kv-a capacities.

Assume for a small transformer 100 kv-a capacity, and for a large transformer one of 4000 kv-a capacity. Now, if the same densities obtain in both copper and iron, namely, if the larger transformer has losses proportional to the increase in output, the losses will be increased $4000 \div 100 = 40$ times, while the area or surface of radiation would only be increased about one-fourth as much as the losses. From this method of comparison it is seen that in order to keep the heating within proper limits it is necessary considerably to increase the size of the transformer, resulting in a cost very much greater than that of a transformer having auxiliary means of cooling. For this reason self-cooled oil-insulated transformers are not generally manufactured in sizes above 2000 kv-a. For sizes up to about 500 kv-a, the tank is single corrugated and above this size compound corrugations are used to obtain the necessary radiating surface.

Hot air tends to flow upward, so that, in providing for station ventilation, it is essential that the inlet of the cool air should be low down and the outlet somewhere near the roof, the inflow and outflow of air being well distributed about the station.

Self-cooled oil-insulated transformers of large size should be given good ventilation or else the life of the transformer will be shortened. The first indication of increased temperature will be darkening of the oil and a slight deposit on the inside surfaces of the transformer. Once this deposit begins to form the tendency is quickened because of the decreased efficiency of heat dissipation from the transformer.

In this type of transformer the only remedy where the oil has thickened to a considerable extent and a deposit accumulated, is thoroughly to clean the transformer by scraping off the deposit and working it out with oil under high pressure.

Transformers Cooled by Forced Current of Air.—This type of transformer is commonly called the "air blast," and may be wound for any desired voltage not exceeding 40,000.

Air-blast transformers are cooled by a blast of air furnished by a blower. The blower may deliver air directly into a chamber over which the transformer is located, or if it is more convenient, the blower may be located at a distance from the transformer, feeding into a conduit which leads to the air chamber. The blower is usually direct connected to an induction motor, though it may be driven by other means. One blower generally supplies a number of transformers in the same station, and the transformers are usually spaced above an air chamber, in which a pressure is maintained slightly above that of the surrounding air. The air for cooling the iron passes from the lower housing selected to suit the transformer capacity. When the efficiency of an air-blast transformer is known, an approximate estimate of the amount of air required can be made by allowing 150 cu. ft. of air per minute for each kilowatt lost. For the most satisfactory operation, the velocity of the air in the chamber should be as low as possible, and should never exceed 500 ft. per minute. That is, the cross-section of the chamber in square feet should at least be equal to the number representing the total volume of air required per minute by the transformer, divided by 500. The power required to drive the blower for furnishing air to the transformers is so small as to be practically negligible, amounting in most cases to only a fraction of 1 per cent. of the capacity of the transformers.

The three-phase, air-blast shell-type transformer, when delta connected, has the same advantage as three single-phase transformers of the same total rating that is, by disconnecting and short-circuiting both windings of a defective phase, the transformer can be operated temporarily at two-thirds, or thereabout, of the total capacity from the two remaining windings.

Coming under this heading of transformers cooled by forced

current of air, there exist two methods, viz., oil-insulated transformers cooled by means of an air-blast at the outside of the tank, and those in which the air is forced directly between the coils, and through ducts in the laminations.

The forced air-cooled transformer may be of the shell type or core type, but preferably of the former for large or moderate sizes.

The question of air-blast against oil-cooled transformers has been settled in practice long ago in favor of the oil-cooled type. Some of the chief advantages claimed are the additional safety due to the presence of the oil round the windings, and the exclusion of a forced current of air and consequently exclusion of dirt and dust from all parts of the windings.

Three-phase transformers require larger air chambers than single-phase transformers of the same total capacity. The temperature of the out-going air compared with the temperature of the in-going air is the best indication whether sufficient air is passing through the transformers, but in general, and on the basis of 25° C., the best results are obtained when the temperature of the incoming air is not greater than this value. Depending on the temperature of the surrounding air or entering air, the out-going air will leave the transformer greater or smaller as the case may be. Also, depending on the design, the difference in temperature of the supply of air and the air leaving the transformer will vary between 12° and 20° C.

The insulation of air-blast transformers must be impervious to moisture, and must have superior strength and durability. It must also permit the ready discharge of the heat generated in the windings, as otherwise the transformer temperature may reach a value high enough to endanger the life of the insulation. In building such a moisture-proof insulation, the coils are dried at a temperature above the boiling point of water, by a vacuum process which thoroughly removes all moisture. After a treatment with a special insulating material, they are placed in drying ovens, where the insulating coating becomes hard and strong. Then the coils are taped with an overlapped covering of linen and again treated and dried, there being several repetitions of the process, depending on the voltage of the transformer. The insulating materials are so uniformly applied and the varnish so carefully compounded that the completed insulation on the coils is able to withstand potentials two or three times greater than the same thickness of the best insulating oil.

Transformers Cooled by Forced Current of Water.—This type of transformer is usually called "oil-insulated, water-cooled." Inside the cast-iron tank and extending below the surface of the oil, are coils of seamless brass tubing through which the cooling water circulates. These coils are furnished with valves for regulating the flow of water, and the proper adjustment having once been made, the transformer will run indefinitely with practically no attention. Another method of cooling is by drawing off the oil, cooling it, and pumping it back, the operation being continuous. In the design of oil-insulated, water-cooled transformers, interior ventilation is provided by oil passages between the coils, and in the iron. These secure an even distribution of heat and a uniformity of temperature throughout the transformer. Without good oil circulation, transformers of large size may reach an internal temperature greatly in excess of that of the external surface in contact with the oil. As a means of securing the best regulation, oil insulation is of immense advantage inasmuch as it permits close spacing of the primary and secondary windings. It effects great economy of space, and its fluidity and freedom from deterioration greatly assist in solving the difficult problems of transformer insulation. Its good qualities come into play with remarkable advantage in building high-potential transformers.

Water-cooling coils are made of seamless tubing capable of withstanding a pressure of from 150 to 250 pounds per square inch.

Transformers Cooled by a Combination Method.—Transformers cooled by this method require the service of a pump for circulating the oil. The oil is forced upward through spaces left around and between the coils, overflows at the top, and passes down over the outside of the iron laminations. With such a scheme transformers can be built of much larger capacities than the largest existing water-cooled transformers of the ordinary type, without such increase in size as to show prohibitive cost and to necessitate transportation of the transformers in parts for erection at the place of installation. The force system allows the circulation of the oil to be increased to any extent, thereby producing a rapid and positive circulation which greatly increases the cooling efficiency of the fluid. Moreover, this method of oil circulation ensures such uniform and positive cooling that much higher indicated temperatures may safely be permitted in transformers operating at moderate overloads.

With ample capacity provided in oil- and water-circulating pumps, the transformer can without danger be called upon to carry extreme overloads under emergency conditions. Transformers of the forced-oil type have recently been built for a normal capacity of 7500 kilowatts, and are actually capable of carrying 10,000 kilowatts continuously at a safe temperature.

Of the very large modern designs of forced-oil type transformers two methods of cooling are employed. The old method was to place the cooling apparatus outside of the tank, and the new method is to place the cooling apparatus inside the transformer tank, the external in this case not being employed. Not unlike the water-cooled type the cooling coil is placed inside of the tank except that it reaches much lower down into the tank than the water-cooled type. A cylindrical or elliptical metal casing, depending on the form of tank, separates this coil from the oil-chamber with the exception of one or two openings at the bottom. The oil is pumped out at the top and into the space enclosing the cooling coil; the static head caused by the resulting difference in level greatly increases the natural oil circulation through the coils and core. This method of cooling is extremely simple in design and is as flexible as any water-cooling system. It is not possible as in the older system of cooling to communicate one transformer trouble to another due to moisture or a breakdown of any kind; it is free to be changed about without interfering with any other transformer in the station, and has several advantages in case of fire. On all very large capacity transformers it is customary to circulate sufficient water (in addition to forced-oil circulation) to dissipate the heat with a rise in temperature of the water of about 10° C. If the temperature of the incoming water is 15° C., that of the out-going should be 25° C. This usually requires about one-third of a gallon per minute per kilowatt loss in the transformers. The rate of flow of the cooling oil through the various coils and core is generally about 25 feet per minute, or somewhere between 15 and 30 feet per minute.

It is well known that high-voltage large-power transformers cannot run continuously, even at no-load, without the cooling medium, since the iron loss alone cannot be taken care of by natural cooling. A device is now in common use whereby an alarm is given the operator when the water has ceased to flow through the cooling coils and also when the temperature has risen above a certain given value in the transformer. This outfit is shown in Fig. 115 and consists of a water relay or balance which is actuated by a volume of water in such a manner that if the water slacks off or ceases to flow, it will light the incandescent lamp shown. The bell alarm is so arranged that it will operate as soon as the temperature of the transformer, as indicated by the thermometer, reaches a certain limit.

In the water-cooled transformer, the heat generated by losses is disposed of as above mentioned, and the arrangement is so

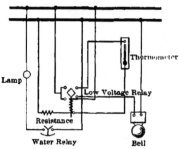

FIG. 115.—Water, thermometer, and bell danger indicator for large high-voltage power transformers.

effective that but very little heat is dissipated from the tank and consequently no advantage is derived from the use of corrugated tanks. There are, however, installations of transformers where a satisfactory supply of cooling water is not available at all periods, and also cases where the meter rate of cooling water is excessive, in which cases a tank with corrugations may be used. The usual design of a tank with corrugations provides for approximately 70 per cent. of total capacity without cooling water. Its cost is, in general, about 10 to 15 per cent. greater than a standard boiler-iron tank. This corrugated type has been used to advantage in very cold climates where water-freezing difficulties are common.

Fig. 116 shows the result of cutting off the water supply of a water-cooled transformer. A five-hour duration has resulted in a temperature rise of 35° C. (60 − 25 = 35° C.), while a two and one-quarter hour duration shows an increase in temperature of 17° C. (57 − 40 = 17° C.).

The cooling coil of this type of transformer is sometimes

METHODS OF COOLING TRANSFORMERS

coated on the outside with a deposit from the oil while the inside is lined by impurities in the cooling water.

A good method of cleaning the inside of cooling coils is to pour equal parts of hydrochloric acid and commercially pure water into the coil. After the solution has been standing for about one hour, flush the coils out thoroughly with clean water.

When deposit has accumulated on the outside of the cooling coil, it is necessary to remove it from the tank for cleansing. The deposit can be wiped or scraped off.

Sometimes moisture is condensed at the top of this type transformer when located in a damp atmosphere. To avoid this, the transformer should be kept warm; that is, the temperature of the oil should never go below 10° C. In some cases a form of "breather" has been used, which consists of a vessel of chloride of calcium, so arranged as to allow the water taken out of the air to drain off without mixing with the air that is going in on account of the contraction of the transformer oil in cooling. An indication of condensation of moisture is the accumulation of rust on the underside of the transformer cover at the top.

On all large high-voltage transformer installations provision should be made for continuous sampling and filtration of the oil in any transformer without removing the unit from service. This is usually done by means of valves at the bottom and top of transformer tank and withdrawing the oil and filtering it by forcing it, at a 200-lb. pressure, through a series of about twenty-five filter sections, each containing five 8-in. by 8-in. filter papers, making a total thickness of almost 0.75 in. of paper. The paper filters the oil and removes all moisture, returning it to the tank dry and clean. The capacity of oil-containing tanks should not be less than the oil capacity of any one transformer, but preferably slightly greater in capacity than the largest transformer in the station. About four hours should be sufficient to filter the entire contents of any transformer and on this basis the capacity of a filter equipment may be provided for.

TEMPERATURE DISTRIBUTION

Core-type (vertical cylindrical coils, turns of the conductor being in a horizontal plane).—With this type of winding the ends of the coils will be somewhat cooler, owing to the heat which passes out at those points. If the temperature of the oil adjacent to the top portion of the coils is much higher than at the bottom, there will be a tendency to transmit heat downward in the coil. The heat transmitted will be small, but the thermal resistance is high as compared with the temperature difference in this direction. The most important result will be that the temperature of the top portion of the coil will be almost as much higher than the temperature of adjacent oil as that of the

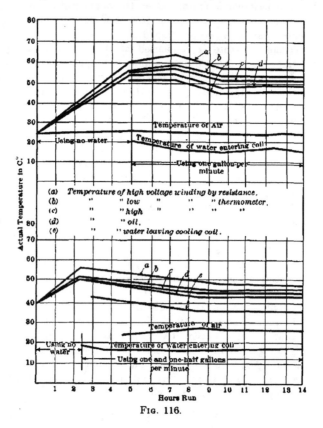

Fig. 116.

bottom portion is above the oil adjacent to it, assuming, of course, that the equivalent thermal resistance from the coil to the oil is practically uniform throughout the length of the duct. The total average temperature of the coil is therefore related not to the surface temperature at the bottom of the

METHODS OF COOLING TRANSFORMERS 127

coil but to a temperature which is average for the entire surface and which may be considerably higher than that at the bottom; also, the maximum temperature at the top of the coils may be considerably higher than the maximum average surface temperature, and should be figured from the surface temperature near the top of the coil.

The difficult problem with natural circulation of the oil is: correct distribution of temperature through the oil from the solid surface of the coils to the surface of the tank, or cooling coil. For a given velocity the equivalent thermal resistance is constant, the temperature drop from coils to oil being directly proportional to the watts per sq. in. discharged from the coils. With forced-oil circulation (returning the oil to the ducts at a definite temperature) the whole problem of cooling is much simplified, the only lacking element being a definite knowledge of the relation between the equivalent thermal resistance at the surface of the coils and the velocity of the oil. With natural circulation the oil flow is caused by the difference in temperature between the average temperature of the oil inside the duct and that of the outside oil between the level of entrance to and exit from the duct.

The effect of the thickness of the duct upon temperature rise will be very different for *forced-oil* circulation than for *natural* circulation. With forced-oil circulation a thin duct will be better than a thick one, since the resulting higher velocity of the oil will give lower temperature drop from the surface of the coils to the oil.

With natural circulation of the oil a thick duct will give practically the same conditions as to temperature rise as that obtained on an external surface, the rise being less than for a thin duct.

Thinning the duct does not cause as great an increase in temperature rise as might be expected, up to a certain point, since, though the temperature rise of the oil while passing through will be greater, this temperature rise tends to produce a higher velocity, and hence to cause a smaller drop from the coils into the oil, as well as to reduce the net temperature rise in the oil itself.

The discharge of heat into the duct from both sides, as compared with its discharge from one side only, is an important matter in connection with cooling. It is found that with a duct of a given thickness, if the heat is discharged into it from both sides, at a given density in watts per sq. in., both the temperature rise of the oil while passing through the duct, and the temperature rise of the coils above adjacent oil, will be smaller than if the heat is discharged into one side of the duct only, at the same density. Thus twice the heat is carried away by the duct, with a smaller temperature rise. This smaller temperature rise of the oil while passing through the duct, though absorbing twice the heat, indicates that the velocity of flow is more than double, and this accounts for the reduced drop from the coils into the oil. The great difference in velocity in the two cases is accounted for by the friction on the side of the duct where no heat is discharged, which is much greater than when heat is being discharged.

Shell type (vertical oil-ducts between flat coils).—With this type, in general, the larger insulation is not in the path of heat flow, and the insulating covering is thin. The equivalent thermal resistance is uniformly distributed throughout the length of the duct, since the rate of oil flow is the same throughout, but this resistance will change with changes in the rate of heat discharge because the velocity of the oil will be different. If the heat generated in any part of the surface which is opposite, the oil would receive heat at a uniform rate throughout its passage through the duct, and the difference between the temperature of the coil and that of the oil would be the same at the top as at the bottom. The temperature at the top part of the coil would be as much greater than its temperature at the bottom as the temperature of the oil leaving the duct is greater than its temperature at entrance. This would result in the passage of a considerable portion of heat downward through the copper, which is a good conductor of heat, the effect being that more heat is discharged per sq. in. from the bottom part of the coil than from the top.

The temperature rise of the oil in its passage through the duct is more rapid in the bottom portion of the duct than at the top, and the temperature drop from the oil is also greater in the bottom portion of the duct than at the top. The temperature gradient in the copper from the top of the coil to the bottom is thus reduced, giving a more uniform temperature, as well as a lower average temperature. On the other hand, though the temperature of the oil where it leaves the duct would be the same, if its velocity were the same, since it absorbs the same total heat, yet its average temperature throughout the duct will be greater on account of the larger proportion of heat which it receives near the bottom. This will result in an increase in the velocity of circulation, which tends to reduce both the temperature rise of the oil in the duct and the temperature drop from the coil into the oil, both of these actions affecting further reduction in the temperature of the coils.

Disc-shape Coils (horizontal position with horizontal ducts between).—These coils may be wound either in single turn layers, or with several turns per layer. In the former case practically all the heat will be thrown out into the horizontal ducts, but in the latter the inner and outer layers will discharge best through layer insulation in the inner and outer cylindrical surfaces. A large portion of the heat will find an easier passage out through the horizontal oil ducts than from layer to layer in the coils. The relative amounts passing out through the two paths will depend upon the circulation of the oil. If the oil is stagnant in the horizontal ducts, it reaches a temperature where it ceases to absorb heat.

It is quite evident that more solid matter of the hydrocarbons will deposit at those points of the maximum temperature; that is, at those points commonly known as "hot-spots." If 80 per cent. of the transformer operates at a temperature of 40° C. and the remaining 20 per cent. at 80° C., the transformer is badly designed and the weakest part of the insulation is at the point of the maximum temperature rise. Therefore, a knowledge of the distribution of temperature throughout a transformer, and of the various things which effect this distribution, is important from the standpoint of all those who operate transformers; and, if we are to avoid serious trouble with transformers, trouble with oil, and oil depositing which necessitates frequent cleaning out of the transformers, it is essential there should be no "hot-spots."

Standard heating guarantees for self-oil-cooled, air-blast and water-cooled transformers are given in the following table.

The oil-cooled transformer measurements are based on the surrounding air temperature of 25° C. normal condition of ventilation, and a barometric pressure of 760 mm. of mercury.

The air-blast transformer measurements are based on an ingoing air temperature of 20° C. and a barometric pressure of 760 mm. of mercury.

TABLE II.—HEATING GUARANTEES FOR OIL-COOLED, AIR-BLAST AND WATER-COOLED TRANSFORMERS

Type of transformer	Temperature rise in degrees C.		
	Full-load continuously	25 per cent. overload for 2 hours	50 per cent. overload for 2 hours
Oil-cooled..........	40	55	60
Air-blast..........	windings 40, core 55	windings 55, core 55	60
Water-cooled.......	40	55	60

The water-cooled transformer measurements are based on a normal supply of ingoing water at a temperature of 15° C.

All corrections for variations in the three above types of transformers are made by changing the observed rise of temperature by 0.5 per cent. for each degree Centigrade temperature variation, or for each 5 mm. deviation in barometric pressure.

TRANSFORMER OIL.

As the subjects of treating transformer oil and the properties of oil are so broad, and have been treated in a thorough manner by other writers, only a few important notes are given here.

The most important characteristics of transformer oil which interest the operating engineer are summed up in the following table.

TABLE III

Characteristics	Quality (A)	Quality (B)
Flash temperature................	188° C.	133° C.
Burning temperature.............	210° C.	146° C.
Freezing temperature............	−10° C.	−16° C.
Viscosity	100 to 105	40 to 42
Specific gravity at 15.5° C........	0.868	0.850
Color of oil.....................	Dark amber	Similar to water

Usually oil is received abroad testing less than 30,000 volts per 0.2 in., but before it is placed into the transformer it is brought up to a test at least 30,000 volts per 0.2 in. for transformers designed for an operating voltage of 44,000 volts and under; not less than a 40,000 volt test per 0.2 in. is required for transformer oil used in transformer operating above 44,000 volts.

At a temperature somewhat below the fire test or burning temperature shown above, the oil begins to give off vapors which, as they come from the surface of the oil, may be ignited in little flashes or puffs of flame, but the oil itself will not support combustion until it has reached the temperature of the fire test as shown above. The lowest temperature at which these ignitable vapors are given off is called the flash point. The difference between flash point and burning temperature test varies considerably in different oils and the actual location of the points themselves varies somewhat according to the method used in their determination. A high flash point and a high fire test are very desirable in insulating oils in order that the fire risk attendent on their use may be reduced to a minimum. Viscosity and flash point vary together, that is to say, an oil having a high flash point, compared with another oil, will probably also be high in viscosity. For all transformers that depend entirely upon oil for dissipating the heat as in the oil-filled self-cooled type, a relatively high flash point is of the utmost importance.

For oil-filled water-cooled transformers it is customary to use another grade of oil than that used in the self-cooled type, the oil operating at a lower average temperature, consequently a high flash is not of the utmost importance. There are several grades of mineral seal oil with flash points varying from 130° C. used in water-cooled transformers.

A very small quantity of water in transformer oil will lower its insulation to a marked degree; moisture to the extent of 0.06 per cent. reduces the dielectric strength of the oil to about 50 per cent. of the value when it is free from moisture. The most satisfactory method of testing insulating oils for the presence of water is to measure the break-down voltage required to force a spark through a given gap between two brass balls immersed in a sample of the oil. Free oil from moisture should have a break-down voltage of at least 25,000 volts between brass knobs 0.5 inch in diameter and separated by a 0.15 in. space.

Often it is desired to determine the insulating qualities of an oil when there are no high voltage testing transformer and apparatus available for making a test. When this is the case, a very good idea of the insulating properties of the oil can be obtained by testing for the presence of water with anhydrous copper sulphate. To prepare the anhydrous copper sulphate, heat some copper sulphate crystals (blue-vitriol) on top of a hot stove. The heat will drive out the water of crystallization, leaving as a residue a white powder, which is known as anhydrous copper sulphate.

Fill a test tube with a sample of the oil, add a small quantity of the anhydrous copper sulphate, and shake well. If there is any moisture present in the oil, it will combine with the anhydrous copper sulphate, forming a distinctly blue solution. This test is quite delicate and a very small quantity of moisture can be detected by it. If this test does not show the presence of water it is quite safe to assume that the insulating properties of the oil are fairly high.

When obtaining a sample of oil for testing, always get the sample of oil from the bottom of the tin or barrel, because, as water is heavier than oil, the maximum quantity of water will always be found at the bottom. To obtain a sample from the bottom, use a long glass tube of small diameter, hold the thumb tightly over one end and plunge it to the bottom of the barrel. Remove the thumb letting the air escape, then press the thumb tightly over the end of the tube and withdraw it with the sample of oil.

The real importance of the light oil known to the trade as mineral seal oil is that it tends to decrease the deposit thrown down on the bottom of the transformer tank, cooling-coils and the transformer coils themselves. The (A) class referred to here is a dark-colored oil having a specific gravity somewhat higher than the mineral oil, its flash and fire points are much higher but when subjected to continued heating it throws down a deposit tending to clog up the oil ducts of the transformer and impede the circulation of the oil. So far as is known this deposit is caused entirely by the effect of the heat on the oil. It can be removed by filtering through a bed of lime and then a sand bed, or preferably by using one of the oil-drying outfits now on the market. The (B) class oil is light and practically white, and is usually referred to as mineral seal oil; it has rather low flash and fire points but does not throw down a deposit when subjected to long and continuous heating. The usual life of transformer oils depends altogether on the thoroughness with which the oil is protected against absorption of moisture, and, when heavy oils are used, the temperature at which the transformers are operated, the higher the operating temperature the more rapid are thin oils affected. It is the general practice to use the mineral seal oil in water-cooled transformers because of its tendency to keep down the temperature and also because it is practically free from the slimy deposit referred to above. With the use of class (A) oil the deposit would, besides accumulating and probably clogging the oil-ducts, close in around the cooling-coils causing a consequent increase in the temperature of the oil and this in turn would decrease the efficiency of the transformer. No specific information can be given as to the length of time throughout which an oil may be in use continuously; it might last five years or it might last only five months, but under ordinary service conditions the oil for high-voltage transformers should be good for at least 18 months. Several good oil drying and purifying outfits are now in common use, the principal elements of this modern outfit being an electric oven, a pump and strainer, filter-press and blotting paper. The interior of the oven is provided with rods for supporting and separating the blotting paper to facilitate rapid and thorough drying. A thermometer is attached to the oven and a switch is provided for regulation of the temperature. For drying and filtering oil, five layers of 0.025-in. blotting paper are used between sections of the press. The separation of this filtering material is of the greatest importance. Special care must be exercised in drying the blotting paper, which should be suspended from the rods in such a way that air is accessible to both sides of each sheet. The blotting paper should be dried at least 24 hours at a temperature not over 85° C., and then put into a tank of dry oil the instant it is removed from the oven and before it is cooled, as exposure to normal air for a few minutes is sufficient to neutralize the drying. It should not come into contact with the hands because of the danger of absorbing perspiration. A small quantity of anhydrous calcium chloride placed in the oven will take up the moisture in the air and quicken the process. A higher temperature than that given above might scorch the blotting paper or impair its mechanical strength. As the paper is somewhat weakened by saturating with oil, it should be carefully handled after removal from the bath. A tank of suitable size for the paper should be filled with dry, clean oil and the paper should be submerged in the oil. The paper should be carefully suspended in the tank, the bottom edge of the paper being kept 3 or 4 in. from the bottom. The oil level should be at least 2 in. above the top of the paper. A strainer is provided to prevent anything of appreciable size from entering and injuring the pump. It is easily accessible and should be cleaned occasionally. The rating in gallons per minute is usually for average conditions in filtering clean heavy oil or dirty light oil. The best oil temperature for filtering is between 25 and 75° C. In the installation of this outfit the pump should be placed so that oil falls by gravity and as fast as the pump will take it. With clean oil the pressure and volume will remain nearly constant, the volume being nearly proportional to the pressure, but with dirty oil the pressure will increase very rapidly. With dirty oil and paper the volume increases much more slowly than the pressure and there is little gained by an increase in pressure over 75 pounds per square inch. In general, class (B) oil is filtered twice as fast as the class (A) oil, and the greater capacity is obtained by frequent renewal of paper, the total capacity per day depending largely on the time consumed in the operation. In placing paper in the press, care should be taken to have the holes in the paper corresponding with those in the plates. Oil is admitted at any pressure not over 100 pounds per square inch. The pressure will at first be very low, gradually increasing as the paper clogs with dirt. It is found that after three replacements of the paper, the dielectric strength of the oil is apt to fall off, and that it is best to discard the full charge of paper and begin again. The amount of oil which can be filtered through one set of papers depends entirely on the quality and temperature, hot oil being filtered with great facility because of its low viscosity. By this process oil may be dried to withstand a puncture test from 40,000 to 60,000 volts with a standard spark gap consisting of two 0.5 in. diameter discs spaced 2/10 in. apart. With oil of an average quality as regards moisture and foreign matter one treatment will usually remove all sediment and bring the puncture voltage to 40,000 volts or more. Oil which has been damaged by overheating from continuous overload or bad burn-out may be treated in this purifying outfit,

METHODS OF COOLING TRANSFORMERS

and the sediment removed although the oil will still be darker in color than it was originally. If the oil is thickened to a slimy nature, it will quicken the operation to heat it to 75° C. just before running it through the press.

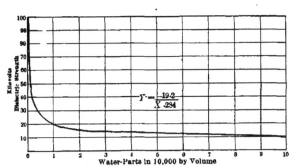

FIG. 117.—Curve showing the effects of water in oil.

The curve Fig. 117 very clearly shows the serious effects of water in amounts less than 0.010 per cent. It shows that the water present in the oil must not exceed 0.001 per cent. in order to obtain a dielectric strength of 40,000 volts in the standard test (0.2 in. between 0.5 in. disc).

CHAPTER IX

CONSTRUCTION, INSTALLATION AND OPERATION OF LARGE TRANSFORMERS

There are various types of transformers on the market differing so much in design that it is difficult to tell exactly whether they are of one or the other design (shell or core type); in fact, what some manufacturers call a shell transformer others call a core-type transformer. However, the design and construction of transformers referred to here are strictly of the core and shell types respectively, of high-voltage large-capacity design, and of American manufacture.

Transformers are always sent from the factory as completely assembled in their tanks as their size and transportation facilities in the countries they have to pass through will warrant. When they are sent disassembled, which is usually the case if they are for very high voltages and large capacity, the tanks are usually protected for shipment abroad but left unprotected for home shipment; the coils are carefully packed in weatherproof boxing, and the core if of the shell type is packed in strong wooden boxes of moderate size (in a loose state), the core of the core type being shipped already assembled each leg being packed in one box and the end-laminations in separate boxes. Whether the transformer is built up at its destination or sent already assembled, it should be thoroughly inspected before being permanently put into the tank. If the transformer is sent from the factory in its tank, which is very seldom done, it should be removed and thoroughly inspected and cleaned before giving it a "heat-run."

Let us deal first with the core-type transformer which consists essentially of two or three cores and yokes which when assembled form a complete magnetic circuit. These cores and yokes are made up of laminated stampings which vary from 0.010 to 0.025 in. in thickness according to the frequency of the system on which they are to be used and the different manufacturers. The laminations are insulated from each other by a coat of varnish or paper to limit the flow of eddy currents.

It is shown in Fig. 118 that there are three cores of equal cross-

CONSTRUCTION OF LARGE TRANSFORMERS 137

section joined by a top and bottom yoke of the same cross-section as the cores, and that upon each core are placed the low and high voltage windings for one phase. The low and high voltage windings are connected so that the fluxes in the cores are 120 electrical degrees apart, making their vector sum equal to zero at any instant.

The usual designs of core-type transformers made by the best manufacturers have a uniform distribution of dielectric flux between high and low voltage windings, excepting at the ends of

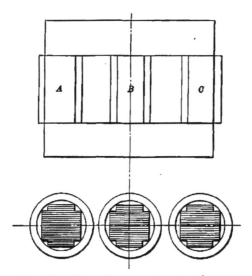

Fig. 118.—Three-phase core-type transformer.

the long cylinders where the dielectric flux will be greater and its distribution irregular.

As already stated, the core-type transformer has its laminations shipped already assembled, wrapped in insulating material of horn-fiber and bound with strong binding tape which serves as a binding to keep the laminations of the assembled sections intact. The different sections are assembled on wooden pins of the size of the holes in the laminations; first, in one direction, and then this end in the opposite direction, alternating spaces thus being left for assembling the end laminations. The approximate number of laminations per inch required in building up these laminated sections may be determined from the information that the iron solid, would be about 90 per cent. of the height of the built up laminations, 36 and 64 laminations per inch being about the number of laminations required for the two standard thicknesses, 0.025 and 0.10 in. respectively. The required number of laminations are built in an insulating channel piece, and on the top of the channel piece and the pile of laminations another channel piece is placed, the whole being pressed down to dimensions and the channel pieces stuck together with shellac under the influence of pressure and heat. The various sections are then assembled with wooden pins to hold them together, these remaining in the holes permanently, and the assembled sections clamped to dimensions and finally wound with a layer of strong binding tape of half-lap, which should not extend beyond the beginning of the spaces for the end-laminations.

The core is now built up, the next process being to insert the end laminations at the bottom, their number corresponding to the spaces left for them, making them so fit into the spaces as to form butt-joints. These laminations are assembled while the cores or "legs" are resting in an horizontal position. While in this position the bottom clamp with its insulation is fastened over the end laminations and the whole raised by the help of another clamp and cross-bars to a vertical position, with open ends upward. The clamping bolts are then placed in a loose state, reliance being put on the bolts (which hold the bottom clamps in position) to keep the cores in a vertical position during the assembly of coil-supports and coils.

The coils, which are of a cylinder form, are raised by means of a stout tape and slipped over the cores. All of the coils which connect together at the bottom should be connected immediately after the coils are in position. Between the cylinder low-voltage coils and the high-voltage coils is placed a cylinder shaped insulating separator, the separator being held in position by means of spacing strips of wood. All connections between outside coils can be conveniently made before the spacing strips are inserted between the high and low-voltage windings, when the coils may be easily turned to such positions as will leave the coil connections as distant as possible from the side of the steel tank. The top connections should be made after all the coils of the high-voltage winding are in position. A pressboard insulating piece or casing is finally placed over the whole

CONSTRUCTION OF LARGE TRANSFORMERS

assembly of coils and tied around with tape. The connections or taps and the terminal leads are brought out at the top and supported in a similar manner to those mentioned in the assembly of shell-type transformers.

The shell-type water-cooled transformers and forced oil-cooled transformers are built in larger sizes than the core type transformers, the former size having been built in 6000 kv-a single-phase units and 14,000 kv-a three-phase units.

The shell-type transformers shown in Fig. 119 consist of three single-phase transformers placed in one tank, the laminated cores being constructed to form a single structure. The reduction of iron for the magnetic circuit amounts to from 10 to 20 per cent. of that used in three single-phase transformers placed side by side.

In this type it is difficult to insulate the large number of edges and sharp corners exposed between adjacent high and low voltage windings and between windings and core. At these points the dielectric density is very great, and it is much more difficult to insulate them than the ends of the cylinder coils of a core-type transformer, and much more insulation is required and consequently a larger space-factor. In fact, a 60,000 volt—2000 kv-a—25 cycle transformer of the core type will have a space-factor of the windings (ratio between the total section of copper conductors to the total available winding space) of about 28 per cent., whereas, in the shell type of the design shown in Fig. 119 the space-factor is only about 17 per cent. The result of this increased space-factor in the winding of a shell type transformer is lower efficiency and worse regulation, and consequently a heavier and more expensive transformer for a given capacity and efficiency. It is sometimes recommended to use graded insulation in high-voltage transformers between the high-voltage and low-voltage windings or between the windings and core because of the unequal distribution of the dielectric flux. In all properly designed transformers for high voltages, the only change in the grade of insulation is at the ends of the windings, this being the only grading considered necessary.

In assembling the coils of a shell-type transformer, this being the first thing to be done in the assembly of this type, care should be taken to eliminate dirt and dust, and the coils at all times should be kept clean and dry. In unboxing the coils, each one should be wiped off with a dry cloth and stacked in the right order of assembly. The assembly of coils is begun in an horizontal plane, and stacked one above the other as shown in Fig. 121, the outer press-board insulation piece being set on two wooden-horses, correctly spaced, depending on the size of the transformer to be assembled. The first coil, taken from the top

Fig. 119.—Three-phase shell-type transformer.

of the stack is placed in position with its inner edges insulated by means of channel-shaped insulation pieces. Each coil before it is placed in the assembly has the same shaped insulation pieces on its outer edges; insulation separators are arranged

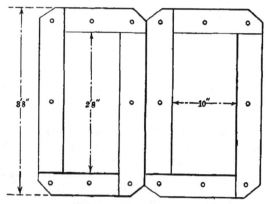

Fig. 120.—Single-phase transformer iron assembly.

in symmetrical order for both the low and high-voltage coils with wooden filling blocks and channel pieces set at the inside top and bottom of coils. The assembly of these coils if of large size can easily be lowered into position.

CONSTRUCTION OF LARGE TRANSFORMERS

The iron laminations must be laid with great care, steel being used to keep the alignment, on which the laminations butt. (See Fig. 120).

In connecting the coils together, all the soldering is best done as the assembly progresses, also all taping of the connections, since the short stub connections are only accessible at this stage of erection. Before commencing to solder, a cloth should be spread over the ends of coils to prevent splashing solder on them which might get inside of the coils and ultimately cause a burn-out. During the assembly of coils, great care should be taken to keep them and the insulation separator in alignment. To facilitate this, all the insulation separators are slit at their four corners and a long round strip of wood is threaded through as the coils are being built up. After the assembly of coils and the placing of the outside top insulating piece or collar position, the whole is clamped down to dimensions. While the end clamps are holding down the coils to dimensions, strong cloth tape is wound around the coils, between the two clamps, under considerable tension, the ends of tape being finally secured by sewing them down, after which the whole is painted with a black, insulating air-drying varnish.

Getting the assembled coils from the horizontal position in which they rest to the vertical position necessary for the assembly of iron and completion of transformer, requires the greatest care especially where transformers are of a larger size than 2000 kv-a. This is accomplished by means of blocking the space, inside of coils, leaving sufficient space in the center of coil-space or a lifting piece. The rope or cable for lifting should be slid over the rounded ends of the lifting piece, care being taken that the wooden spacer is sufficiently high about the coils to prevent striking the sling when swinging to vertical position.

The bottom end frame or core support is now ready to be set into position where it is desired to build up the iron, which is arranged in a similar manner. The coils should be exactly vertical. Lapping of the laminations should be avoided, otherwise difficulty will be experienced in getting in all of them. Raw-hide mallets should be used for driving the laminations into line; or in case these mallets are not available, hard wood pieces pressed against the laminations may be hammered. During the building up of the laminations they should be pressed down two or three times, depending on

142   STATIONARY TRANSFORMERS

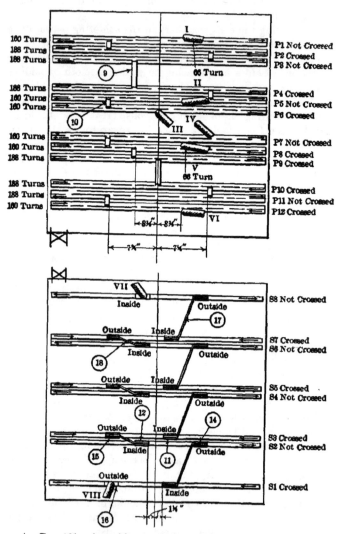

Fig. 121.—Assembly of coils for a shell-type transformer.

the size of transformer being built. For this clamping, the top iron frame is usually lowered and forced down with especially made clamps or the regular clamping bolts may be used if the threading on them is sufficient to lower the clamp to the required dimensions; these clamps or bolts may be left in position over night if any difficulty is found in getting in the laminations. In the case of transformers without a top piece or frame, an especially constructed rigid frame of wood may be used. The laminations should be built up to such a height as will permit the core-plate to be forced into position under considerable driving. It is always a difficult matter to put in all of the laminations that come from the factory as special facilities for pressing are available there. After the laminations have been clamped down, the low- and high-voltage leads should be supported and insulated in a right manner; care being taken in supporting the leads to see that they are properly spaced, for if they are placed too close together a short-circuit might occur. At the cast-iron frames, the insulator bushings used for this purpose are held by metal supports to the frame, and at the point where the leads pass through the bushings, cement is used to hold the leads into position. Both the high and low voltage winding end-leads connect to the terminal-board which is located above the assembled coils and core. In the case of oil-cooled and water-cooled transformers all the leads are brought out at the top; usually air-blast transformers have their leads terminate at the base or part at the base and part at the top of assembled coils.

In lowering transformers into their tanks, care must be taken to get the bottom cross-bar, into which the tie-bolts are screwed into the correct position, which requires that the transformers should be properly centered in the tank.

Of the two types of transformers, the core type is the easiest one to assemble, and a full description of one particular method of assembly of this type might well be said to cover practically every manufacture, whereas the shell type varies in many ways and is a difficult piece of apparatus to assemble, especially in the larger sizes.

Air-blast transformers are regularly built in capacities up to 4000 kv-a and for voltages as high as 33,000 volts. The efficiency of this type of transformer, in good designs and with the required amount of air pressure, is sometimes better than the oil-filled water-cooled type of the same capacity and voltage. Its general design is very much like the ordinary shell-type transformer with the exception of a few modifications in the iron assembly where certain air-spaces are left open for the circulation of air. The air space area in these transformers is considerably in excess of the actual area required for the pressure of air specified for cooling. For this reason dampers are provided to regulate the air so that, where a number of transformers are involved, each transformer will receive its portion. The air always enters the transformer at the bottom and divides into separate paths, flowing upward through the coils and ducts controlled by the dampers at top of the transformer casing, and through the core ducts controlled by a damper at the side of the casing.

This type of transformer is always shipped already assembled so that in the larger sizes great care is necessary in handling them. For shipment abroad the larger sizes would, of course, have to be disassembled, but as this is of rare occurrence the above holds for practically all sizes.

These transformers are placed over an air-chamber made of brick or concrete the sides being made smooth to minimize friction and eddy currents of air. From the blower to the chamber should be as free as possible of angles, and where these occur, they should be well rounded off. Sufficient space should be allowed for the location of the high- and low-voltage leads, and necessary repairs and inspection. Three-phase transformers have larger air-chambers than single-phase transformers of the same aggregate capacity. The temperature of the out-going air compared with the temperature of the incoming air is the best indication whether sufficient air is passing through the transformers; if there is not more than 20° C. difference, the supply of air will be found sufficient. As transformers are generally designed on the basis of 25° C. the best results are obtained when the temperature of the in-coming air is not greater than this value.

Installation of Transformers.—In the installation of high voltage transformers of the self-cooled oil-filled, water-cooled oil-filled, forced-oil-cooled oil-filled, and the air-blast types, the following points of importance should be born in mind:

(a) In generating and receiving stations the transformers should be so situated that a burn-out of any coil, a boiling over

CONSTRUCTION OF LARGE TRANSFORMERS

of the oil, or burning of the oil in any unit will not interfere with the continuity of service.

(b) In generating and receiving stations the transformers should be so located that the high voltage wiring from transformers to bus-bars is reduced to a minimum.

(c) The transformer tanks, which are, of course, made of a metallic or non-combustible material, should be permantly and effectively grounded, preferably to the ground cables to which the station lightning arresters are connected.

(d) Sufficient working space should be allowed around each unit to facilitate the making of repairs and for necessary inspection.

(e) During the entire process of assembly of transformers of low or high voltage, the best and most careful workmanship is of utmost importance.

(f) (This might be considered as the last process in the installation of transformers but by no means the least important.) Extra special knowledge and care is necessary on the part of all those whose duty it is to dry out transformers—the difficulty is not in drying out the coils, as is usually supposed, but *the drying of the whole insulation surrounding them and the core.* No matter what the factor of safety the transformer has been built for it avails little in the case of carelessness or neglect to dry out the transformer properly.

Before transformers leave the factory they are given a high-voltage test, the standard being, to apply twice the rated voltage between the high- and low-voltage windings, the latter being connected to the iron core. The main object of applying this test which induces twice the rated voltage to one of the windings, is to determine whether the various portions of the coils are properly insulated from each other. It is now believed that the greater causes of failure in high-voltage transformers are punctures between turns and not between the high and low-voltage windings.

To install properly and place in good working order high-voltage power transformers is quite as important as their design, since upon this depends the life of the transformer. All transformers of high voltage should be thoroughly dried out on arriving from the factory, and all transformers which show evidence of being unduly moist, or that they have been subjected to conditions that would cause them to be unduly moist, should be taken special care of in the drying process.

Testing Cooling Coils.—Before high-voltage transformers are put into operation they are subject to a "heat-run," and in the case of transformers with cooling coils, the coils are made subject to a pressure test. These coils must be assembled before the heat run can be made. If the coils show evidence of rough usage, such as heavy indentations and disarrangement of layers, the coils should be given the usual tests to determine whether a break has resulted. The method for testing for leaks is to fill the cooling coil full of water, establish a pressure of 80 to 100 lb. per square inch, disconnect the source of pressure, holding the water in the cooling coil by means of a valve, and note whether the pressure gauge between the valve and the cooling coil maintains the indication throughout a period of about one hour. Care should be taken that no air is left in the cooling coil in filling it with water. In removing the source of pressure it is preferable to disconnect entirely from the cooling coil, in order both to make sure that the source of pressure is entirely removed and to note whether the lowering of the pressure indicated by the gauge connected to cooling coil, is due to leakage through the cooling coil valve or to leakage through a hole in the cooling coil. If the gauge indicates a lowering of pressure in the cooling coil, it should be inspected throughout its entire length until the hole is discovered. The water will gradually form at the hole and begin to drip. After the cooling coil is filled with water, a small air-pump may be used for giving the required pressure, in case there is not a satisfactory water source for obtaining the pressure. As the test is only to determine whether the cooling coil has a leak in it, it will in no case be necessary to establish a greater pressure than 100 lb. per square inch. Some engineers prefer to submerge the cooling coil in a liquid, under an air pressure of 80 to 100 lb. per square inch for a period of about one hour, and note the bubbles rise to the surface of the liquid.

Drying-out Transformers.—Several methods exist for drying out high-voltage transformers, the best being considered as follows:

1. Short-circuit either the high or low-voltage winding and admit sufficient current to raise the temperature of the windings to approximately 80° C. The amount of heat necessary to effect this temperature will range between one-third and one-fifth of the full-load current, depending on the room temperature

and design of transformer. The impedance volts necessary to give the specified range in current, vary from 0.4 per cent., to 1.5 per cent. of the rated voltage of the winding to which the impedance voltage is applied. In any case, the current admitted must be so regulated that the temperature of the windings does not exceed the 80° C. limit.

The temperature of the transformer windings may be determined by the increase in resistance, or, if facilities for this method are not available, the bulb of a *spirit* thermometer may be placed in direct contact with the low-voltage winding at the top. Low-voltage winding is specified for the reason that to place the bulb of the thermometer in contact with the high-voltage winding may not give the temperature of the coils; the insulating pieces set around the coils of the high-voltage winding being built up on the copper under the tape to such a height as to prevent the thermometer recording the temperature of the copper. The bulb of the thermometer should be placed down between the low-voltage coils as far as possible. Mercury thermometers must never be used for this purpose because of their liability to break. The drying process should be carried on while the transformer is out of its tank in order to give as good a circulation of air as is possible under the conditions.

The following table is considered to be within safe limits for carrying on the drying process although discretion must be used as in the case of an unduly moist transformer and where the kw. capacity and voltage enter into consideration. That is to say, a transformer coming under the heading of 44,000 to 70,000 volts, having a kw. capacity of 200 kw. and less, can be taken as safe if the heat-run period is only carried on for 60 hours instead of 72, or a transformer coming under the heading of 22,000 to 33,000 volts, having a kw. capacity of 200 kw. or less may be considered safe if the heat-run period is carried on for only 24 hours instead of 48 hours; assuming all transformers in normal condition:

It is impossible to give anything but an approximate estimate of the number of hours necessary to dry out a transformer of a given size and voltage. Much will depend on the condition of the transformer when it is received from the factory, whether in an unduly moist condition or dry.

(2) This method is to dry the transformer and oil simultaneously under the influence of heat and vacuum, the transformer

TABLE IV—APPROXIMATE HOURS NECESSARY IN DRYING OUT HIGH-VOLTAGE POWER TRANSFORMERS

Voltage of system	Hours of heat-run	Kilowatt capacity
22,000 to 33,000	38	200 and above
22,000 to 33,000	48	500 to 1000
22,000 to 33,000	60	1000 to 2000
22,000 to 33,000	72	2000 and above
33,000 to 44,000	52	200 to 500
33,000 to 44,000	64	500 to 1000
33,000 to 44,000	72	1000 to 2000
33,000 to 44,000	84	2000 and above
44,000 to 66,000	64	200 to 500
44,000 to 66,000	72	500 to 1000
44,000 to 66,000	84	1000 to 2000
44,000 to 66,000	92	2000 and above
66,000 to 88,000	84	500 to 1000
66,000 to 88,000	96	1000 to 2000
66,000 to 88,000	118	2000 and above
88,000 to 110,000	96	500 to 1000
88,000 to 110,000	118	1000 to 2000
88,000 to 110,000	130	2000 and above
110,000 to 145,000	140	2000 and above

being dried inside of its tank. The tank is first made vacuum tight, this being, in the majority of cases a difficult task to do and is only accomplished after considerable time has elapsed with the vacuum pump under operation, by closing the holes indicated by the whistling noise of the entering air. The leaks are stopped by using putty, which should be fairly stiff in order to keep it from being drawn into the tank. If the puttying is done a day or two before the drying process is begun, thus giving the putty a chance to harden, it will be found much easier to obtain the required vacuum.

One of the transformer windings is short-circuited as in method (1), although the actual temperature in this case is allowed to reach 90° C. instead of 80° C., and the temperature is determined by the increase in resistance. The temperature of the oil should be maintained at approximately 80° C. during the drying process. When starting the heat-run it is found advantageous to bring the temperature up quickly, and to do this, full-load current might be given until the approximate temperature is reached, after which it should be reduced to the specified value.

CONSTRUCTION OF LARGE TRANSFORMERS

In addition to heating by electric current a certain amount of heat should be applied under the bottom of the base of the transformer. The most satisfactory method of applying heat to the base is to use grid resistances supplied with sufficient current to maintain the grids at full red heat. The grids should be distributed under the base so as to make the heating fairly general, and not confined to one portion of the surface. In case some other method of heating the base is used, extreme care should be taken that the supply of the heat does not become too intense, otherwise the oil may be injured. The idea of supplying heat to the base is to maintain uniform temperature of the oil throughout the transformer structure at a uniform temperature of 80° C. It is found that the temperature of the windings reaches 90° C. considerably in advance of the oil's reaching 80° C.; and, for this reason, it is necessary either to disconnect the current occasionally or to reduce it to a small percentage of full-load current. The base heating should be relied on to maintain the oil at a temperature of 80° C. as long as it will, which may be almost constantly, provided a sufficient quantity of heat is applied. These specified limits of hours referred to above are for the actual time the process must be carried on after the oil has reached a temperature of 80° C. and after a vacuum of 20 in. has been established, and does not refer to the time necessary to reach the 80° C. point and 20 in. of vacuum.

When electric current is not available, steam at a low pressure may be used for heating, the steam being admitted through the cooling coil. Also, steam may be used for the base heating; in which case the entire bottom surface of the base should be subjected to the heat of the steam. Care should be taken in admitting steam through the cooling coils that the temperature of the oil does not exceed the limit. This method of applying heat at the base is not recommended, principally because the steam condenses on all parts of the transformer tank.

(3) This method of drying transformers requires the circulation of heated air through the transformer coils and core while it is in the tank. The source of heated air should be connected to the base valve and the top cover of tank be partly removed. The temperature of the air inside of the tank should be maintained at approximately 80° C., and the process should be carried on under this temperature for a period of three days for units of moderate size, the same discretion being used as mentioned in

methods (1) and (2). The temperature of the heated air as it enters the transformer should not exceed 100° C. This method of drying transformers is especially adapted to localities where no electric current is available.

The oil may be dried by the vacuum method mentioned in (2), or by blowing heated air through it, referred to in the above method. Where the vacuum method is used, the tank must be filled to within a few inches of the top, so that the cover may be kept sufficiently warm to prevent condensation of moisture. In case the transformer tank without its transformer is used for this purpose, it is sometimes necessary to put temporary bracing inside of the tank to prevent collapsing under vacuum; this does not refer to the tank of cylinder form. A 12-hour run under a temperature of 80° C. with not less than 20 in. of vacuum should be quite sufficient to dry transformer oil. All large installations are provided with tanks for this purpose, the tanks being of the cylinder form.

The necessary heat for bringing the temperature of the oil up to 80° C. may be obtained by placing a steam coil or an electric heater in the bottom of the tank. Assuming that the tank in which the oil is being dried, will radiate approximately 0.25 watts per square inch, the amount of electric energy required to maintain the oil at the specified temperature may quite easily be estimated. The electric heater should be about double the size estimated for the purpose of shortening the time necessary to reach the desired temperature. Whether a steam coil or an electric heater is used it must be placed directly on the bottom of the tank as it is necessary to maintain the oil temperature about uniform throughout. In case steam is used, its pressure should not be greater than 10 lb. per square inch.

The same tank may be used for drying oil by means of forced circulation of air. In this case it is necessary to run the piping from the valve in the base of transformer up above the oil level, and then down to the air pump, the top of the tank having an adjustable opening for permitting the air to circulate. The oil must be heated to a temperature of approximately 100° C., and the process continued until the oil becomes dry as determined by test.

Comparison of Shell and Core-type Transformers.—Transformers of any type should not be selected at random but only after careful investigation of design, reliability and simplicity to repair.

In general the shell type transformer is a difficult piece of apparatus to repair in case of a break-down; the difficulty increases in almost direct proportion with increase in capacity, and in the larger sizes it becomes advisable to send for a transformer man from the factory to do repairs. This disadvantage has been and is to-day considered the only cause of a large number of power companies operating their lines at high voltages choosing the core-type transformer.

Experienced transmission engineers never fail to realize the severe conditions to which transformers are subjected to in practice, and whenever they ask manufacturers for transformers to connect to their high-voltage lines, seldom fail to go thoroughly into the factor of insulation, which, to them, means continuity and uninterrupted service. It is well-known that the insulation of a high-voltage transformer is subject to very severe potential strains, some of which, are:

(a) Sudden increase in generator voltage.
(b) Sudden increase in line voltage from local causes.
(c) Direct and indirect lightning discharges.
(d) Ground on one of the lines—depending on the connection.
(e) Internal or external arcing grounds.
(f) Line surges, etc.

Reliable data taken from a number of power companies operating long-distance high-voltage transmission lines, show, that the shell-type transformer has been more reliable than the core type for high-voltage service.

With such large sizes as there are operating to-day, this class of apparatus can very well be considered as one of the most important, if not the most important piece of apparatus connected to a transmission system, and its reliability to satisfactorily operate for long periods of time, after it has once been put into service, is looked on with much interest and wonder from every side. To think, as some do, that once a transformer has been put into successful operation it will continue to operate indefinitely in a satisfactory manner without any attention is a wrong idea. It requires attention, and must be given attention or else it will not give good service.

Late modifications in the grading of insulation on the end turns of the core-type transformer has given it a better standing, and it is now considered to give better service and can be depended upon in this regard equally as well as the shell-type transformer.

Some of the most important advantages and disadvantages of these two general types might be summed up as follows:

Advantages in Favor of the Shell-type.—Greater radiating surface of coils and core resulting in a lower temperature in all parts of transformer. This point has an important bearing on the insulation; the life of the transformer depending on the strength of the insulation of the hottest part.

Interlacing of coils resulting in lower reactance voltages, hence closer regulation.

Mechanically stronger and better able to withstand the electro-magnetic stresses. As the electro-magnetic stresses are proportional to the square of the current, a short-circuit of many times the normal full-load current will produce abnormal strain in the transformer.

Satisfactory series-parallel operation. This often being necessary on large transmission systems.

Advantages in Favor of the Core Type.—Easier to repair.

Disadvantages of the Shell Type.—Difficult to remove a coil.

Disadvantages of the Core Type.—With low-voltage winding designed for 22,000 volts and above, the amount of insulation next to the core means a larger mean turn of winding; the temperature and $I^2 R$ loss being increased thereby.

Radiating surface on the low-voltage winding very poor, resulting in higher temperatures. It is a disadvantage if, say, 90 per cent. of the transformer operates at a temperature of 50° C. and the remaining 10 per cent. at 80° C. as this point is the weakest link in the insulation.

The arrangement of coils (concentric) results in poorer regulation and higher reactance voltages.

Less mechanical bracing because of its design and form.

Not possible to operate a three-phase (delta-delta) transformer in case one winding becomes damaged.

It is obviously true that equally good results can be obtained with either the core-type or shell-type construction, but the design of one or the other would depart from the regular standard expressed above if equal performances and reliability of operation for equal conditions of load, etc., are desired. The difference is slight but nevertheless in favor of the shell-type construction, particularly from the operating point of view, with the exception of repairs, should a break-down occur.

The shell-type transformer is cheaper than the core-type with *dear copper* space (large copper space factor *and ordinary iron*). And likewise, the core-type transformer is cheaper than the shell-type with relatively *cheap copper space* (low copper space factor *and alloyed iron*). In other words, shell- and core-type transformers cost as nearly as possible equal amounts WHEN equal volumes of copper and iron spaces are *equally expensive*.

Operation of Transformers.—The general idea is that transformers do not require any care or attention but that they will operate quite satisfactorily after having been put in service and no further attention is necessary. This is, however, not the way to get good results, although in some fortunate cases it might apply and has applied. Many losses of large power transformers have been recorded resulting from the cessation of the cooling medium, all of which could have been saved if proper care had been given to them. Hourly temperature reading is the best indication of anything wrong in this direction. As is well known, high-voltage transformers designed to operate with some form of cooling medium cannot run continuously, even at no load, without the cooling medium, since the iron loss alone cannot be taken care of by natural cooling. In case the circulation has been stopped by any cause, the transformer may be operated until the coils at the top of transformer in the case of an air-blast, or until the oil in case of a water-cooled transformer, reaches an actual temperature of 80° C. This temperature limit under ordinary conditions will permit the transformer to continue delivering power for about three hours; a very close watch must be kept of the temperature and the transformer must be taken out of service as soon as it reaches this limit.

The efficiency of a transformer is usually considered to be its most important feature by the majority of central station engineers and managers operating distribution systems. By transmission engineers this factor is not considered to be the most important, the most important being the insulation and mechanical strength of the transformer, and consequently its reliability. The efficiency is no doubt an important feature and should not be neglected in the choice of a transformer but it cannot be considered as the most important feature of a large high-voltage power transformer. The writer believes the right order of importance to be:

1. Reliability, or ability to supply continuous and uninterrupted service.
2. Safety, or a condition conforming with safety to life and property.
3. Efficiency, or a condition met with after proper allowance has been made to conform with 1 and 2.

Reliability as referred to here means many things both internal and external. It might be stated, but not generally, that efficiency is the principal point in discussing transformers from their operating point of view. Where large high-voltage power transformers are concerned, efficiency can be said to take the third place of importance, and probably second place where low-voltage city distribution transformers are concerned.

Excepting cases of lightning and roasting of coils due to constant overload, the low-voltage transformer (as used for city lighting and motor service) is free from harm. Not so with the large high-voltage power transformers such as we are using at the present day in connection with long-distance transmission systems. Causes of failure are numerous; that is to say, causes that would and sometimes do bring about burn-outs of coils, moving of both coils and iron, and complete failure of a transformer.

Use and Value of Reactance.—A transformer may be absolutely reliable electrically but be weak mechanically which might bring about its wreck.

We know the short-circuit stresses are inversely proportional to the leakage reactance of the transformer, therefore on large systems of large power the use of high reactance or, more correctly, high automatic reactance, is required that will vary with the current. Several interesting methods are from time to time suggested and some of them experimented upon, but, to keep the current down to even 15 times normal current on a "dead" short-circuit and brace the transformer to take care of such a shock and be certain that the transformer is quite strong enough to take care of other severe mechanical stresses we need more experience. Some of the methods tried and in their experimental stage, are:

(a) Use of reactance, internal or external, or both.
(b) Use of resistance in the neutral of grounded systems.

CONSTRUCTION OF LARGE TRANSFORMERS 155

(c) Use of induction generators at the generating stations in place of the ordinary synchronous machine.

(d) Strengthening of the transformers themselves.

Experience has already demonstrated that due to excessive short-circuit current a limiting reactance of some form or other must be provided for large power transformers operating on systems of large kilowatt capacity, its design being either stationary or rotary; that is to say, an inductive reactance or an induction generator. With a stationary reactance it is quite possible to arrange to have it automatically switched into circuit, or switched out of circuit as the case might warrant when a short-circuit occurs. This would mean that it is used only when required and is always out of circuit under normal operating conditions.

The ability of a modern high-voltage power transformer to withstand short-circuits is of far greater importance than good regulation. Modern practice does indicate larger reactance in both generators and transformers when operating in connection with long-distance high-voltage systems, and regulation of 6 per cent. or worse is not considered *very* bad. Where ordinary city distribution transformers are used regulation is of another order. With our larger systems, larger stations and larger transformers, reactance of a certain given value in addition to that originally put into the system, station and transformers is virtually important and may be necessary if a limit is to be provided to check the enormous amount of power that can be developed in a short circuit.

Causes of excessive current and consequent mechanical strains are: short-circuits between lines, short-circuits between transformer terminals and leads and bus-bars, and grounds on star-connected systems with grounded neutral and other systems with grounded neutral.

In order to arrive at the mechanical stresses occuring in a transformer due to a short-circuit it is simply necessary to use certain expressions in terms of the terminal voltage, short-circuit current, and the distance between primary and secondary coils. For example:—Take a 5000-volt, 5000-kw., 25-cycle transformer, full-load current in the primary being 577 amperes and measured reactance of the windings, is 2.9 per cent.

If (as is usually done) the reactance of the windings is given in per cent. of the impressed voltage, the short-circuit voltage will be equal to the full-load current divided by the percentage reactance, or:

$$I^0 = \frac{I}{X_e} = \frac{577}{0.029} = 20,000 \text{ amperes} \qquad (18)$$

which represents an amount in excess of 30 times normal full-load current.

If (f) is the force in grams produced between primary and secondary windings, and (l) the distance between their magnetic centers, the mechanical work done in moving one set of coils through the distance (l) against the force (f) would be:

$$F = f.g.l. \times 10^7 \text{ joules} \qquad (19)$$

At short-circuit very little magnetic flux passes through the secondary coils of a transformer, but if the system is sufficiently large to maintain constant voltage at the terminals of the transformer during a period of short-circuit, full magnetic flux passes through the primary coils. Such a condition can never exist sufficient to maintain a constant voltage at the terminals of transformers located at the end of long-distance transmission lines, but for comparatively short distances and on systems of practically unlimited power behind the transformers (sufficient to roast them) and of close regulation, it is possible to get results that will do so much damage as to wreck them entirely. Thus a transformer with a 2.9 per cent. reactance would give a short-circuit current at constant voltage of 30 times full-load current, and one with a reactance of 2.3 per cent. would give 40 times full-load current, while one with 4 per cent. reactance would produce only a short-circuit current of 25 times. This, then, certainly demonstrates that more reactance in the transformer circuit for better protection is required, and that the reactance should be designed proportional to the current so as to be effective.

The terminal voltage at the transformer during short-circuit is taken from the leakage inductance of the transformer, therefore:

$$\frac{810 \, E \, I^0}{f.l.} \, g.c.m. = \frac{0.706 \, E \, I^0}{f.l.} \text{ inch lb.} \qquad (20)$$

is the force exerted on the transformer coils and represents the work done in moving the secondary coils until their magnetic

CONSTRUCTION OF LARGE TRANSFORMERS

centers coincide with those of the primary coils (an impossible condition) which would cause zero reactance flux to pass between primary and secondary coils.

Now assuming that the transformer in question has *three* primary coils between *four* secondary coils, and the distance between the magnetic centers of the adjacent coils, or half-coils, is *three* inches. The force exerted on such a transformer and its respective coils would be something like:

$$F = \frac{810 \, E \, I^\circ}{f.l.} g = 4263 \times 10^5 g$$

$$= 940,000 \text{ lb.} = 426 \text{ tons.}$$

This force is exerted between the *six* faces of the three primary coils and the corresponding faces of the secondary coils, and on every coil face is exerted the force of:

$$\frac{F}{6} = \frac{426}{6} = 71 \text{ tons}$$

If the distance between adjacent coils had been 1.7 inches the force exerted on the transformers would have been 1,750,000 lb., and if the distance had been 4.3 inches instead of three inches, the force exerted on the transformer would have been under 400 tons, or about 65 tons for each coil; this is the average force, which varies between 0 and 130 tons.

In the design of such a reactance coil the following formula is used:

$$E = \frac{4.44 \, f \phi N}{10^8} \qquad (1)$$

in which (f) is the frequency, (N) the number of turns, (ϕ) the flux enclosed by the conductor. The flux produced by a coil without an iron core being:

$$\phi = \frac{INd}{k} \qquad (21)$$

where (N) is the number of turns, (d) the inside diameter, (k) a constant which equals $0.28 + 0.125 - \left(\frac{l}{D}\right)^1$, and I the current in the coil.

The tendency of leakage flux is not uniform throughout the width of the coil of a transformer, but is greater at the center of

---

[1] (l) is the length and D the mean diameter of the solenoid.

the coil (if the coil be imbedded in iron). If, then, a short-circuit should occur, the coil will have a tendency to buckle, and should it not be sufficiently strong to overcome this tendency, and its equilibrium be disturbed, the forces of conductor upon conductor will not lie in the same plane, and hence are liable to break the insulation tape bindings, and heap up on each other at the point of most intense density.

The shell-type transformer coils have the tendency to twist at their outer corners, and the vertical portions of the coils to form into cable.

The core-type transformer coils have the tendency to be forced upward or downward into the iron core (the heaviest current coils being, in most instances, forced out of position). If, however, the centers of the primary and secondary are exactly coincident the entire force would be exerted in a horizontal direction, and there would be no tendency for any of the coils to move vertically. Whether the primary or secondary is forced up depends upon which coil has its center line above the center line of the other.

It is only recently that the power-limiting capabilities of reactance have come to the fore. This has largely been due to the marked movement toward consolidation and concentration in the central station industry resulting in unified systems of gigantic proportions, the loads upon which may fluctuate suddenly through a wide range or, still worse, short-circuits on such high powered systems may give rise to rushes of current the volume of which was hitherto unknown in previous systems. In the last few years we have come to larger and larger systems, and consequently greater difficulties of operation. The concentration of power for economical reasons in these huge power plants, the dependence for vast industrial enterprises and for our ever increasing transpoitation systems, as well as for lighting and industrial power from ten, twenty or thrity substations distributed over vast areas and supplying large cities, make it absolutely essential that they shall be protected against disturbance, and that every possible precaution should be taken, which experience or ingenuity can provide, against irregularity in operation, because if these huge transformer systems are going to be subjected to disturbance and interruption of service other factors affecting their regulation and efficiency must be sacrificed to gain this end, if need be.

CONSTRUCTION OF LARGE TRANSFORMERS

The advantages of iron and air reactance coils in a power station are many, provided the former type is worked at a sufficiently low magnetic density so that it will not become saturated at the maximum peak of a short-circuit current, and provided the latter type has not too strong an external magnetic leakage, and that both types are used only when needed. They are effective in protecting transformers against surges, lightning, and short-circuits.

At 110,000 volts and over a phenomenon makes itself felt, especially in large transformers, which is negligible at lower voltages; the distributed capacity of the high-voltage transformer winding. At lower voltages, the transformer capacity (c) is negligible, and the transformer thus is an inductive apparatus, and as such is free from all high-frequency disturbances, such as traveling waves, impulses, stationary oscillations, etc. High-frequency currents cannot enter the transformer, but produce high voltage between the end turns, orotection against which is given by the high insulation of the end turns of the transformer, and also the external or internal choke coil. At very high voltage the electrostatic capacity of the transformer becomes appreciable, and the high-potential coils of the transformer then represent a circuit containing distributed capacity, inductance, resistance and conductance. In the high-voltage winding of a transformer the inductance is high in value and the capacity low, that is, comparatively speaking lower than the respective constants in a high-voltage transmission line. The result, in general, is that the oscillations are higher in voltage and lower in current, in the former. The danger to which a transformer is exposed by high-frequency disturbances from the line side, is not limited to the end turns only, but damage may be done anywhere inside of the transformer.

A choke coil or reactance between the transmission line and transformer might or might not in this case become a source of danger. It protects the transformer from certain line disturbances but does not protect the transformer itself from disturbances which originate inside its windings; in fact the addition of this choke coil has the tendency to throw back the disturbance and thereby increase the internal voltage and destructiveness.

Recently it has become customary to specify that transformers of large sizes and high voltages must not have less than approximately 5 per cent. reactance for the protection of transformers, switches, generators and all parts of the system against the high mechanical stresses due to excessive currents.

To *increase* the reactance of a given transformer one or several modifications are possible, as, for instance:

(1) Decreasing the dimensions of the windings in the direction in which the leakage flux passes through the wire-space.

(2) Decreasing the number of groups of intermixed primary and secondary coils, the number of turns of each group being correspondingly reduced.

(3) Increasing the *total* number of turns in primary and secondary.

(4) Increasing the length of turns in primary and secondary.

From the viewpoint of safety to the transformer itself by the introduction of higher reactance within the transformer, little practical benefit is derived.

No hard and fast rule can be given for the correct location of reactance. There are several reasons against making it *all* a part of the transformer. Of course, so far as the generating stations are concerned inductive reactance in any form in connection with high systems is important, but where transformers are located at the end of long transmission lines it may or may not be necessary to use it. To make a large transformer of large power with high reactance is not an easy matter, as the general principles of design and the economic utilization of materials obtain for us only low factors, and in order to make a transformer of high reactance we have to increase its cost and its ampere turns, with of course more copper and more winding space, and consequently a larger core and a bigger transformer for the same output. If it is desired to increase the reactance by increasing the space between the primary and secondary windings, the same results are obtained and the efficiency of the transformer is reduced. Placing reactance in the transformer itself is very effective on short-circuit, whereas an external current-limiting reactance will not generally be so effective because it is not on general principles designed for the short-circuit current and its value consequently is about frustrated by magnetic saturation. A reactance is required only during the short-circuit. In that case, why then should an expensive and inefficient transformer be considered when a short-circuit might occur only once in five years; would it not be much better to arrange a reactance external so that its maximum flux on

CONSTRUCTION OF LARGE TRANSFORMERS

short-circuit is about equal to its voltage flux? If a reactance is put in to limit the short-circuit current, the reactance must be there when the short-circuit occurs. All long transmission lines possess magnetic reactance which tends to reduce the voltage so that less reactance will be required for similar transformers located at the end of the line than those at the generating station.

Earthing the Neutral of Transformers.—The chief advantage of resistance in the neutral of a star-connected system is to limit the earth current on short-circuit. To arrive at a close value of resistance that will limit the earth current at *all times* is not easy, in fact in most cases impossible; the ideal condition being where the load and voltage of the system are not disturbed. The two extreme conditions of operation are the insulated system and solid-grounded system. Of these two, experience so far has proved the better to be the grounded system, and whether we ground through resistance or "dead" ground the method depends entirely on the system itself. Both methods will work satisfactory while certain conditions of operation exist and likewise both will vary under one particular condition, it being better or worse in one or the other depending on the kind of disturbance. It seems to appear that the best way to limit the current and disconnect the circuit at the same time, in preference to grounding through resistance, would be to add an automatic reactance in the line side which would come into effect when the line current reached a given value and actuate the oil switches. A broken insulator or ground of any kind on the line, develops a short-circuit which will interrupt the service depending on the exactness of the resistance in the grounded circuit and neutral; generally speaking the service is interrupted whether there be resistance in the grounded neutral or not. It is often stated that with one line down and grounded in an insulated delta system (a non-grounded delta system), an interruption or short-circuit will not occur. This statement is very vague and might lead those who have not had experience with very high-voltage system (voltages above 60,000 volts) to believe that in the majority of cases it is correct. It may be stated that it is not correct for neither the non-grounded star or delta systems operate at these high voltages except in very unusual cases.

The primary or secondary may be put to *earth* or ground through a group of star-connected transformers as shown in Fig. 122, or, if these transformers are not available and a ground connection must be used, a substitute might be made similar to Fig. 123. With this connection grounded as shown, the maximum insulation strain between any of its secondary or primary windings respectively to ground (whichever side might be grounded) will be 87 per cent. of full voltage between terminals, but with the transformer method of grounding and under the same operating conditions the maximum strain will not be greater than 58 per cent. of full voltage between terminals. Grounding the neutral point

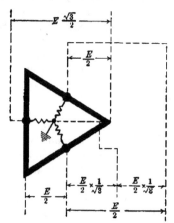

Fig. 122.—Method of grounding an insulated delta system through inductance coils.

of the high-voltage windings of transformers connected to a transmission line will immediately operate the relays or circuit-breakers should one of the line conductors fall to the ground. Grounding the neutral point of the secondary, or low-voltage windings, will not operate the relays unless some special provision has been made, and even so, before the relays or circuit-breakers are actually operated the secondary or low-voltage windings are made subject to very high-voltage stresses. The best method of all is to ground the neutral points of both the high- and low-voltage windings and where delta-delta systems are used ground the windings as shown in Fig. 123. A common condition found in practice is the non-grounded delta-delta system. On systems where the ground connection is used, the character of the ground

CONSTRUCTION OF LARGE TRANSFORMERS 163

connection and the ground itself should be considered at the worst time of the year *only*, inspection and tests at other times of the year being of little or no value and should not enter into the design of the resistor and its earth resistance, and the earth connection itself. Grounding through a limiting resistance gives certain advantages, and grounding solid has the disadvantage of interrupting the service at all times in case of any other ground developing in the metallic circuit. Its use is helpful in reducing mechanical stresses and in largely overcoming dangerous surges set up due to arcing grounds, etc.

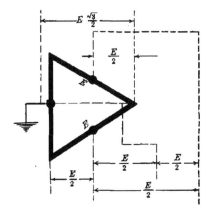

FIG. 123.—Method of direct grounding an insulated delta system.

Earth Connections.—The old method of making earth connections consisted in excavating a large hole, placing an expensive copper plate in this hole surrounding it with a load of coke. At the present time this method is generally considered to be not only a waste of money, but of less efficiency than the multiple pipe-earths. Their resistances are no lower even where perforated and treated with special compounds, their current capacity no greater and their life and constancy no better.

For very small areas such as pipe-earths, the resistance of an earth connection depends greatly upon the exposed area of the metal plate to earth.

A simple contact of a metal conductor with a normally moist earth will give a high resistance of enormously variable values due to the variations in contact.

An ordinary pipe resting on the ground was found to give an average of 2000 ohms. The same pipe driven 6 feet into the ground gave 15 ohms, and the same pipe resting on dry pebble gave several thousands of ohms.

As the pipe penetrates the earth, it is found that each additional foot adds a conductance about proportional to the added length.

The specific resistance of the earth will depend upon what chemicals exist around the metal plate and how much moisture there is present. In a dry sand-bank the resistance is practically infinite. In a salt marsh the specific resistance is extremely low, being about one ohm. Resistances of earth connections will vary greatly even in the same locality.

The engineer is interested mostly in the earth connection in the immediate vicinity of the earth pipe or plate; because in the main body of the earth, the area of cross-section through which there is current, is so enormously great that even if the specific resistance is very high the total resistance becomes negligibly small. If the earth plate should lie in the dry non-conducting stratum of the top layer, it is advisable to get some means of introducing better conductivity, not only in the contact between the plate and the earth, but also between the earth conducting layer deeper down. The best means of accomplishing this is to pour a salt solution around the iron pipe and allow it to percolate down to a good conducting stratum. In order that this solution may not be washed out by the natural filtration of rain water, it is well to leave a considerable quantity of crystal salt around the pipe at the surface, so that rain water flowing through will dissolve the salt and carry it continuously to the lower strata. Salt has the additional value of holding moisture.

Objections have from time to time been made to the use of salt in stating that it would be destructive to the metal of the pipe. Under the usual conditions it is found that the chemical action on an iron pipe is of negligible value. Iron pipe is very cheap, and it would be better practice to use the salt, even if it did destroy the pipe within a period of years which it does not.

If it is desired to decrease the resistance of earth connections, it is necessary to drive earth pipes that are separated by a distance sufficient to keep one out of the dense field of current of the other. The current density in the earth around a pipe-earth drops off approximately as the square of the distance. A good method is to drive multiple pipes at least 6 ft. apart and connect them together, and the resistance will decrease almost in proportion to the number of pipes.

The more salt water placed around a pipe-earth, the less the potential gradient near the pipe. Inversely the drier the earth, the more the concentration of potential near the pipe.

Earth pipes have a certain maximum critical value of current which they will carry continuously without drying out. The application of a high voltage might have so bad an effect that the earth around a pipe will be dried quickly and the earth-plate lose its effectiveness as a ground. As the moisture is boiled out and evaporated at the surface of the pipe, the surrounding moisture in the earth is being supplied, but the vapor generated tends to drive away this moisture from the pipe.

The diameter of the pipe effects the resistance comparatively little. Doubling the diameter of a pipe decreases the resistance by a small percentage.

Switching.—In laying out systems which must of necessity be more or less complicated, the question of continuity of service should be always kept in mind and placed first in importance. At the present time there are in operation many intricate systems with tie-lines between receiving stations and generating stations. This arrangement presents difficulty in providing protection on account of the interconnections and disturbance in any one line may cause an interruption of a large portion of the system.

Systems are sometimes "overrelayed," relays which the operator cannot thoroughly understand being installed. The result is that he oftentimes renders some of them inoperative by plugging with wood to prevent what he considers unnecessary interruptions of service, which may be at the expense of needed protection to the transformers. Before deciding upon the system of connections a careful study should be made to determine the simplest possible arrangement, when taking into account not only the delivery of energy under normal conditions, but also continuity of service under abnormal conditions.

Low-voltage moderate capacity transformer switching is simple. High-voltage power transformers connected to large systems are not quite so simple in their arrangement of switching and their protection from both internal and external short-circuits and other faults.

Figs. 124, 125 and 126 show four different methods for high-voltage transformer protection and their switching. The series relays may be attached directly to the switches which they operate, while those relays energized by means of series transformers and potential transformers (differential relays) are usually

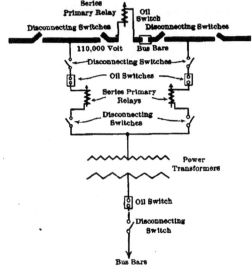

FIG. 124.—Method of switching and protection of transformers.

attached to the switchboard panels located on the operating gallery.

Series relays, or secondary relays (relays energized from the secondaries of series transformers) or both, may be used on high-voltage circuits. The oil-switches marked A, B, C, etc., are operated by means of intermediate low-voltage switches which receive their energy from an auxiliary source of supply or the same source as the case may be; these small switches are located on the main switchboard. The disconnecting switches are of the air-break single-pole type and are usually located at that point where a break in the circuit is most desired, for interchanging, or isolating a circuit, bus-bar, etc.

If a transformer is switched direct on to a high-voltage system

CONSTRUCTION OF LARGE TRANSFORMERS 167

there may be a rush of current equivalent in its suddenness to what usually occurs in high-frequency experiments, and the exact amount of the rush will depend largely on the local capacity of the high-voltage switches and their connections, that is, the distance they are located from the transformers to which they are connected.

The usual shocks to the end turns of transformers due to switching are dependent upon the load, character of the load,

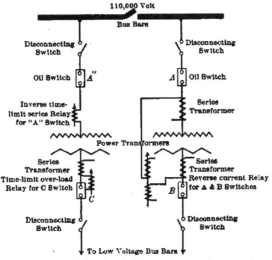

FIG. 125.—Two methods of switching and protecting transformers.

type of switch (oil type or air, and single- or multi-break), the time allowed for opening, and the method of switching. As regards the switching itself, there are several ways in which transformers can be switched on to a line, as for example:

(1) Switching in the transformers on an open line.

(2) Switching non-energized transformers on an energized line.

(3) Switching energized line on to energizied transformers.

(4) Switching in the transformers and afterward raising the voltage at generating station.

(5) Switching under any of the above conditions but with resistance or reactance in series.

(6) Switching on the low-voltage side of transformers.

(7) Switching on the high-voltage side of transformers.

(8) Switching energized transformers on *long transmission lines* with transformers already switched in at the receiving station, but "dead."

(9) Switching in transformers during lightning storms, line disturbances, etc.

(10) Switching in transformers using oil-switches.

(11) Switching in transformers using air-break switches.

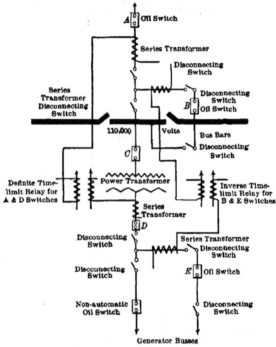

Fig. 126.—Another method of transformer switching and protection.

Most of the methods above shown are bad and may lead to excessive surges being thrown on the windings of the transformers. Next to (4), the best method is to do all switching on the low-voltage side whenever possible using only 3-pole oil-switches in the case of three-phase circuits. The method (4) is generally not possible because of the necessity of lowering

CONSTRUCTION OF LARGE TRANSFORMERS

the voltage of the system. The next preferential method is where a receiving station is "dead" and has to be energized all high-voltage switches being closed first.

A very important point about switching is the time-limit. If it is necessary to open a circuit instantaneously on very large loads, switching can be made large enough to do the work, but where economy of design is required, a time-limit should be used, and, probably, a reactance to limit the flow of current. The time-limit with its relay will therefore take care of that amount of current which it is set for and will permit the switch to open up to the load under which it is able to operate safely. If a reactance is used in connection with switches of this kind, it might be well to provide for some arrangement whereby it can be brought into use only when it is needed or, for instance, when the circuit is actually being opened.

Aside from the above methods of *closing* circuits, there are two conditions of *opening* circuits which should be avoided, as:

(1) Opening a long transmission line under heavy load, on the high-voltage side.

(2) Opening a long transmission line on the high-voltage side with no-load.

Method (1) refers principally to short-circuits, and method (2) to a high-voltage line required to be made "dead." If a live circuit is to be cut out at all it should be done from the low-voltage side no matter what the load conditions are.

High-voltage switches for large power transformers should be of the most substantial mechanical construction and capable of safely breaking a circuit under extreme conditions. For comparatively small installations it is still customary to use expulsion fuse-switches installed inside of a delta connection, that is, one on each lead or 6 for a three-phase group. This is not good practice for the reason that with a blown-out fuse (leaving the delta open) a bad phase distortion may occur. Where connections of this kind are necessary, they should be made through air-break switches and not fused switches.

CHAPTER X

AUTO TRANSFORMERS

The ordinary auto-transformer is a transformer having but one winding. The primary voltage is usually applied across the total winding, or in other words, across the total number of turns, and the secondary circuit is connected between two taps taken off from the same winding, the voltage ratio being equal to the ratio of numbers of turns.

The auto-transformer shown in Fig. 127 has two taps brought out at a and b. Thus the whole or part of the winding may be

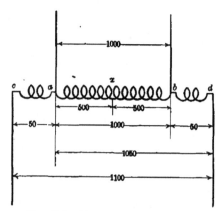

FIG. 127.—Step-up auto-transformer.

used to raise the voltage or lower the voltage simply by changing the connections.

For example, the primary $a\,b$, is wound for 1000 volts, $a\,c$ and $b\,d$ each being wound for 50 volts. As will be seen, by taking a tap out from a and d, the secondary gives $1000 + 50 = 1050$ volts. And by moving to the far end of the winding, a, the voltage may be raised from $1050 + 50 = 1100$ volts. In order to obtain 550 volts all that is necessary is to bring two leads out from x and d, or x and c; the secondary then gives $1050 - 500 = 550$ volts.

AUTO TRANSFORMERS

For pressure regulation auto-transformers are very convenient, being used to some extent for regulating the voltage of transmission lines. They are also used for starting induction motors; and lately they have been used for single-phase railway service, rectifiers, low voltage (5-15-27- . . . . . volts), lighting, etc.

For series incandescent systems a transformer similar to that shown in Fig. 127 may be used. A portion of the winding, $a\,b$, is common to both primary and secondary. The secondary voltage, $c\,d$, is greater than the primary, $a\,b$, by the voltage of the winding, $b\,d$ and $c\,a$. The voltage, $b\,d$ or $c\,a$, is thus added to the primary to form the secondary voltage of the circuit.

By reversing the connections of the winding, $b\,d$ and $c\,a$, however, it may be made to subtract its voltage from the primary,

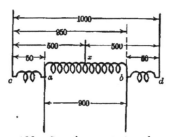

Fig. 128.—Step-down auto-transformer.

$a\,b$; in which case the secondary voltage becomes less than the initial primary voltage; (Fig. 128). Further, by bringing a number of leads from parts of the winding, $b\,d$ or $c\,a$, the secondary voltage may be increased or decreased by successive steps as the different leads are connected to the secondary circuit. For a given transformation of energy, an auto-transformer will be considerably smaller than an ordinary transformer, and consequently its losses will be less and the efficiency higher. The amount of power delivered to the service mains at an increased voltage is very much greater than the power actually transformed from the primary to the secondary of the transformer. In fact, the power actually transformed is equal to the increase of voltage multiplied by the total current delivered; and the output, or actual rating of the transformer is based upon the power transformed.

Example.—The voltage of a long-distance transmission line is to be raised from 40,000 to 45,000 volts, and the maximum current to be handled is 750 amperes. What is the rating of auto-transformer required for this service? and what will be the actual power delivered over the line?

The actual rating of the auto-transformer will be,

$5000 \times 750 = 3750$ kilowatts.

The total power delivered to the line will be,

$45,000 + 750 = 33,750$ kilowatts.

In Fig. 128 the secondary voltage is smaller than the primary. The voltages, $b\,d$ and $c\,a$, are thus subtracted from the primary to form the secondary voltage of the circuit. The auto-transformer may thus act as a step-up or step-down transformer.

The action of an auto-transformer is similar to that of the ordinary transformer, the essential difference between the two lies in the fact that in the transformer the primary and secondary windings are separate and insulated from each other, while in the auto-transformer a portion of the winding is common to both

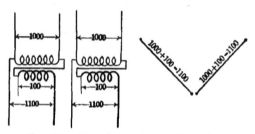

FIG. 129.—Two-phase auto-transformation.

primary and secondary. The primary and secondary currents in both types of transformers are in the opposite direction to each other, and thus in an auto-transformer a portion of the winding carries only the difference between the primary and secondary currents.

In the foregoing explanation of auto-transformation the ordinary transformer will be used instead of the auto-transformer.

Two-phase, four-wire, auto-transformation as shown in Fig. 129, where the secondary winding is made to assist the primary, may be considered as two ordinary single-phase circuits. The ratio of transformation in this case is 10 to 1, therefore we obtain by the connection as shown, a secondary voltage of 1100.

By reversing the secondary connection it is possible for us to get $1000 - 100 = 900$ volts.

AUTO TRANSFORMERS

If we should take one end of the secondary winding and connect it as shown in Fig. 130 we would obtain 50 per cent. of the primary voltage plus 100, which is the total secondary voltage. Then assuming the primary and secondary to have a four-wire, that is to say, two independent single-phase systems, we would have a

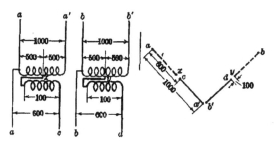

Fig. 130.—Two-phase four-wire auto-transformation.

secondary voltage of 500 plus 100 = 600 volts. The points, x and y, are taps brought out from the middle points of the windings.

Another two-phase auto-transformation is represented in Fig. 131. Both primary and secondary are connected to a three-wire

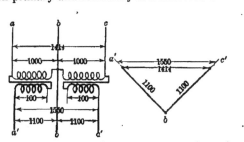

Fig. 131.—Two-phase three-wire auto-transformation.

system from which we obtain a secondary voltage of 1100 between a' b and b c', and 1550 volts between a' and c', or

$$1100 \times 1.41 = 1550 \text{ volts.}$$

A very interesting combination giving a five-wire, two-phase transformation is shown in Fig. 132. From this arrangement it is seen that quite a number of different voltages and phase relations can be obtained, and by simply shifting the connection at x we increase and decrease the resultant voltages.

At $a\,x\,d$ and $c\,x\,d$ the respective phases that constitute the four-phase relation have been changed from 45 degrees to a

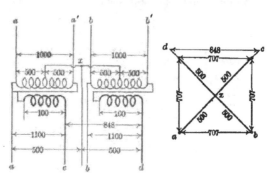

FIG. 132.—Two-phase five-wire auto-tranformation.

slightly higher value, the voltage increasing in proportion to the increase of phase difference.

The three-phase arrangement shown in Fig. 133 is a method of auto-transformation by which we are enabled to supply approxi-

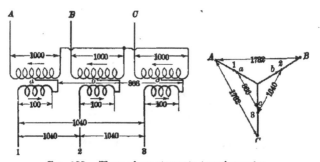

FIG. 133.—Three-phase star auto-transformation.

mately 1040 volts to the secondary mains, 1, 2 and 3, from a 1732-volt primary source of supply, using three transformers with a ratio of 10 to 1, or = 1000 to 100 volts.

Between points $a\,b$, $b\,c$, $a\,c$, we obtain

$$500 \times \sqrt{3} = 866 \text{ volts.}$$

AUTO TRANSFORMERS

Between points 1 2, 2 3, and 1 3, there exists approximately

$$500 + 100 \times \sqrt{3} = 1040 \text{ volts.}$$

The three-phase delta connection shown in Fig. 134, with its three secondary windings left open-circuited, may be used where

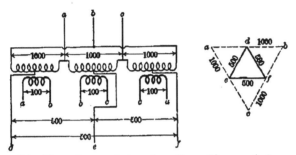

Fig. 134.—Three-phase auto-transformation with secondaries open circuited.

a three-phase 500-volt motor is installed. The secondary windings, if required, may be used at the same time for lighting or power. To obtain a 100-volt lighting service it will be necessary to connect the secondary windings in delta, running a three-wire distribution to the source of supply, $a\,b\,c$. This method of

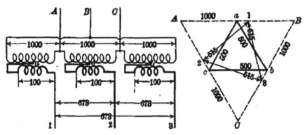

Fig. 135.—Three-phase delta auto-transformation.

connecting transformers is often found useful in places where transformers of correct ratio are not obtainable.

The combination shown in Fig. 135 has its secondary windings connected in circuit with the primary windings. Like Fig. 133 a tap is brought out from the middle of each winding; but instead of leading out to the secondary distribution, it is connected to one end of the secondary winding as shown at 1, 2 and 3; the result of which represents a phase displacement as shown in the vector diagram.

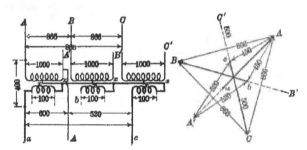

Fig. 136.—Three-phase double-star auto-transformation.

Using the same transformers as in the previous examples, connecting $A\ B\ C$ to a $500 \times \sqrt{3} = 866$ volt supply, it is possible for us to obtain a number of different voltages for the secondary distribution, such as three at 1000, three at 600, three

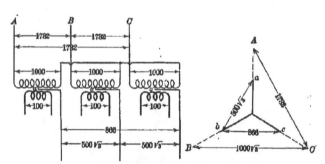

Fig. 137.—Three-phase star auto-transformation with secondaries open-circuited.

at 520, six at 500, three at 173, and three at 100 volts, respectively (Fig. 136). According to the raito of transformation applied at the secondary distribution it is understood that the kilowatt capacity of the transformers will vary.

Another three-phase combination is shown in Fig. 137, where it is shown that the primary windings are connected in star, and

AUTO TRANSFORMERS

the three leads, $A\,B\,C$, are connected to a 1732-volt supply. From the middle of each primary winding a tap is brought out at a b and c.

The secondary voltage across $a\,b$, $b\,c$ and $a\,c$ is $500\times\sqrt{3}=866$ volts. The 100-volt secondary winding may be used for power

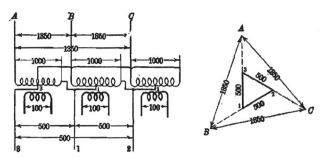

FIG. 138.—Three-phase auto-transformation using primary windings only.

and lighting, single or polyphase, depending upon the size and design of the transformer.

Fig. 138 represents a three-phase transformation, using only the primary windings. One end of each primary winding is connected to the middle point of another primary winding. Three-phase, 1350 volts, impressed on $A\,B\,C$ will give 500 volts on 1-2, 2-3, and 1-3.

CHAPTER XI

CONSTANT-CURRENT TRANSFORMERS AND OPERATION

For operating arc- and incandescent lighting systems from constant-potential, alternating-current mains, the constant-current transformer is frequently used. It is designed to take a nearly constant current at varying angles of lag from constant-potential circuits, and to deliver a constant current from its secondary winding to a receiving circuit of variable resistance.

Thus the transformer operates automatically with respect to the load, making it possible to cut out any number of lamps, from full rated load to zero load, while still maintaining a constant current on the line. The self-regulating characteristic is obtained by constructing the transformer in such a manner that either the primary or secondary coil is balanced through a system of levers against a counterweight, which permits the distance between primary and secondary coil to vary. This automatically increases or decreases the reactance of the circuit in such amount as to hold the current constant irrespective of the load.

For the majority of series incandescent systems the constant-current transformer will be found lower in initial cost and more reliable in service than the reactive coil method, as it combines in one element the advantages of a regulating device and an insulating transformer.

One type of transformer consists of a core of the double magnetic type with three vertical limbs and two flat coils enclosing the central limb. The lower coil, which is fixed, is the primary, while the upper one, or secondary, is carried on a balanced suspension, and is free to move along the central limb of the core.

The repulsion between the fixed and moving windings of the system for a given position is directly proportional to the current in the windings.

For series enclosed, arc-lighting on alternating-current circuits the constant-current transformer is universally used. This type usually consists of a movable secondary and fixed primary windings, surrounded by a laminated iron core. This core and the yokes at the top, bottom and sides, form a double magnetic circuit, as shown in Fig. 139.

The magnetic flux which passes through the primary winding, flows partly through the secondary winding. The secondary winding is made movable and partly counterbalanced by a weight so that an increase in the current causes the secondary to be pushed further away from the primary. The weight is so adjusted as to sustain the coil against the leakage flux, and simply by changing the amount of counterweight the transformer can be adjusted to maintain any desired current.

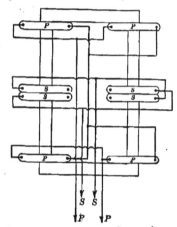

FIG. 139.—Type of constant-current transformer for arc lighting systems.

In small transformers, which have but one movable coil, the counterweight equals the weight of the coil less the electrical repulsion, and a reduction in the counterweight will produce an increase in the current. In large transformers, having two sets of movable coils balanced one against the other, the counterweight serves merely to draw the primary and secondary coils together in opposition to the repulsion effect. In this case, a decrease in the counterweight is followed by a decrease in the current.

The counterweight attachment is made adjustable because the repulsion exerted by a given current in the coils is not the same at all positions of the coils, being greater when the primaries and secondaries are close together and less when the primaries are separated. When the primaries and secondaries are separated by the maximum distance, the effective force tending to draw them together should be less than when they are in full-load position; that is, when the primaries and secondaries are close together.

For capacities of 100 lamps, or less, there is one primary and one secondary coil, the primary being stationary, and the secondary, or constant-current coil is suspended and so balanced by weights that the repulsion between it and the primary changes the distance between them with variations of load, the current in the secondary being kept constant.

For capacities of 100, or more, there are two primary and two secondary coils. A separate circuit of lamps may be operated from each secondary, the two circuits being operated at different currents if desired.

The maximum load of each circuit, when operated separately, will be one-half the total capacity of the transformer. However, when it is necessary to operate the two circuits at unequal loads, the load of one circuit being less, and of the other greater, than one-half the rated capacty of the transformer, the coils may be connected together in the multi-circuit arrangement, which will allow loads up to the total capacity of the transformer to be carried upon one circuit.

For capacities of 250, or more, there may be one or two primary and two secondary coils, two circuits being operated from each secondary, thus giving four circuits from the transformer. It is not necessary that the loads on the two circuits from each coil be balanced, and, if desired, the total load can be carried on one circuit alone, provided the insulation of the line is such as to admit the high voltage which will be introduced. Constant-current transformers are of the air- and oil-cooled type. The air-cooled type is surrounded by a corrugated sheet iron or cast-iron casing with a base and top of cast iron. The oil-cooled type is surrounded by a cast-iron case, providing ample cooling surface. The working parts are immersed in oil, which assists in conducting away the heat.

Constant-current transformers are usually located in stations where electric energy is generated, received or transformed, as, for instance, a receiving station at the end of a long high-voltage line; although in large cities they are, for convenience, located in district sub-stations and close to distribution centers.

They are made to operate on 60- 125-cycle and even 133-cycle systems, and for any reasonable primary voltage. It is customary to furnish a 60- 125-cycle transformer for a 125- 133-cycle system.

The constant-current transformer will maintain constant current even more accurately than the constant potential transformer maintains uniform potential, and its regulation over a range from full-load to one-third its rated capacity will come within 1.5 per cent. if properly adjusted.

The efficiencies of constant-current transformers with a full load of arc lamps vary at 60 cycles from about 96 per cent. for the 100-lamp transformer to about 94.5 per cent. for the 25-lamp transformer.

Another constant-current transformer known as the "edgewise wound" type is fast replacing the older type mentioned above.

From this modern method of construction several advantages are derived, principal among which is the almost absolute impossibility of an internal short-circuit, as the voltage between any two adjacent conductors consists of only the volts per turn of the transformer, or at the most about 10 volts.

The construction of this type is somewhat different. The core is built up of thin laminations of sheet steel and has a center leg of cruciform shape. This form of construction not only tends to support itself, thereby requiring a very thin angle for securely clamping the laminæ and decreasing the eddy current in the clamp, but the form of construction also gives the most economical flux path as well as permitting a smaller diameter coil.

The primary and secondary consist of four concentric edgewise wound coils of double cotton covered rectangular wire. The four sections are assembled together concentrically with wooden spacing strips to maintain at all points an air-duct of sufficient width. Two surfaces of each conductor are therefore exposed to the currents of air passing through the air-ducts, thereby increasing the effective radiating surface of each coil by about threefold. The large radiating surface with the consequent cool running allows a very high current density which permits of less copper per ampere-turn, less weight, less floor space, and, of course, a cheaper transformer for the same kilowatt rating than the older type above mentioned.

182 STATIONARY TRANSFORMERS

Construction of Transformer.—The core of this efficient and most modern type of constant-current transformer is built up of laminations of specially annealed iron, which are sheared to the required length and width. Each sheet or laminæ is treated by coating the surfaces with a species of japan, which serves to reduce materially the eddy current loss in the core. This japan is applied by passing each individual sheet between rolls which are constantly kept moist with the japan. After passing the rolls, the pieces of iron are carried along a travelling table, where they are dried by passing nozzles through which air is blown.

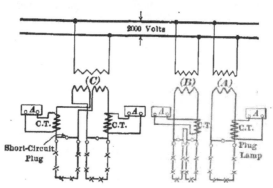

Fig. 140.—Connections for single, or multiple of series arc or incandescent circuits.

Like the core-type constant potential transformer the construction commences with the iron, the iron legs being assembled in a horizontal position. After all the laminations have been stacked together and wrapped with the necessary sheets of "horn-fiber," the whole assembly should be placed in a press under considerable pressure which will reduce the weight of the iron to the required dimensions. In the factory an hydraulic press is generally used, the pressure applied being equivalent to several tons; the temperature is also increased to about 250° F. There are three legs in each transformer—two of rectangular shape, and one of "cruciform" shape which is larger in cross-section than the two rectangular legs—each leg being assembled and handled separately until completed, after which the three are raised vertically and accurately spaced for the placing of end laminations. After the end laminations have been put in and the whole assembly made ready for the placing of coils, the completed iron core is turned upside down.

The form of coil for this modern type constant-current transformer is similar in every respect to a large majority of those used in constant potential power and lighting transformers. It is cylindrical in form and consists of rectangular shaped wires, the width being several times the thickness. All the coils are placed on the center leg which is the one of cruciform dimensions. In all, there are four coils located concentrically with insulating spaces between each coil. For arc and incandescent lighting it is customary to make the lowest voltage coil the movable coil. For use with mercury arc rectifier systems the primary coil is the movable coil, while the secondary coil is the movable coil for series alternating-current lighting.

The fabrication of this class of coil for constant-current transformers is interesting. In the winding of the coil a collapsible cylindrical former is used which has the same inside diameter as the required inside diameter of the coil. The wire is set on edge and wound in this manner around the former. In the winding, the wire is fed through a friction device to give the required tension, the starting end of the wire being clamped to a flanged collar revolving with the winding form. The wire is pressed firmly against the collar by another collar, which loosely fits the winding form and is held stationary; next follows another flanged collar which presses heavily against the stationary collar, thereby forcing the several turns of wire very tightly against one another. The flanged collar thus travels slowly along the winding former, so that for one revolution of the machine, or turning lathe, it travels a distance equal to the insulated thickness of one turn of the wire. After the coil has been finished and taken from the collapsible former it is set into a clamp and baked at a temperature of about 180° F. in a well-ventilated oven, thereby removing all moisture; and, while still hot, is dipped in a tank containing insulating varnish. This heating and dipping is repeated several times until the coil becomes self-sustaining and until the insulation will take up no more varnish. The coil is then wound with tape, each turn overlapping the preceding turns by one-half to one-third its width. Various kinds of tape are used in the insulating of coils, such as cotton tape, which is varnished after applying, varnished cambric tape, which is treated after applying, and mica tape. Mica tape is only used on very high-voltage transformers. Where cotton tape is employed it receives a brushing of the best quality of insulating varnish, is then baked, revarnished and rebaked, this process being repeated several times for each tape.

The regulation of this transformer comes within one-tenth of 1 amp. above or below normal current from full-load to no-load.

Fig. 140 (*A*), (*B*) and (*C*) show three different methods of connecting the secondaries of constant-current transformers for single, or multiple of series arc or incandescent circuits. Method

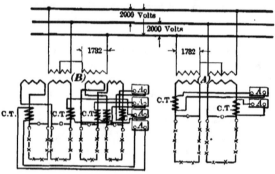

FIG. 141.—Method of operation from three-phase primaries, using two constant current transformers in each case.

(*A*) represents a simple circuit; (*B*) a single secondary winding with two circuits operated in series or singly as desired; (*C*) shows a multi-circuit secondary, each circuit being operated independently.

Fig. 141 (*A*) and (*B*) represents two methods of operating constant-current transformers and series-arc lighting from three-phase primaries. The secondaries of (*A*) show two independent circuits, each circuit being supplied from separate transformers, the primaries of which are wound for 2000 volts and $2000 \times \sqrt{3/2}$ volts respectively. The secondaries of (*B*) also show two independent secondary circuits, each being in two parts as shown.

Before shipment, constant-current transformers are made subject to an insulation test of 10,000 volts between secondary, primary and all parts; also between primary, secondary and all

CONSTANT-CURRENT TRANSFORMERS

parts. The duration of the insulation test is one minute. If however, the primary voltage is above 5000, which is very rarely the case, the insulation test is twice normal voltage.

The modern type of constant-current transformer referred to above, has a guaranteed temperature rise not exceeding 55° C. based on a room temperature of 25° C. If the temperature of the room is greater than 25° C., 0.5 per cent. for each degree difference should be added to the observed rise of temperature; if less, subtracted.

The record tests of a 100-lamp 6.6 amp., air-cooled constant-current transformer are shown below.

CONSTANT-CURRENT TRANSFORMER RECORD TEST
Full-load = 100 lamps — 6.6 amp. — 60 cycles

Test	Lights connected				
	100	90	80	70	60
Core loss in watts..................	1215	1050	892	892	892
Copper loss in watts...............	656	635	611	504	402
Sec. open-circuit voltage............	9500	8660	7950	6940	5930
Pri. current at 2200 voltage.........	29.0	27.3	25	21.8	18.7
(Temperature rise after 12 hours' run 55° C.)					
Efficiency in per cent. at full-load....	96.1	96.6	96.1	95.8	98.5
At 75 per cent. full-load..........	94.3	94.9	94.8	94.5	94.1
At 50 per cent. full-load..........	92.5	92.5	92.4	92.0	91.4
At 33 per cent. full-load..........	89.1	89.2	89.1	88.5	......
Power factor in per cent. at full-load.	72.6	71.8	69.6	69.7	70.2
At 75 per cent. full-load..........	55.2	54.3	52.9	53.0	53.2
At 50 per cent. full-load..........	37.8	37.2	36.2	36.3	36.4
At 33 per cent. full-load..........	26.1	25.7	25.0	25.0	......

When operating a load of 6.6 amp. lamps plus 7.5 per cent. line loss, the voltage and power factor at the lamp are:

Volts per lamp + line loss = 83 per cent.
Power factor of lamp = 84 per cent.

CHAPTER XII

SERIES TRANSFORMERS AND THEIR OPERATION

The characteristics of the series transformer are not very generally known. It is used in connection with alternating-current ammeters and wattmeters where the voltage of the circuit is so high as to render it unsafe to connect the instrument directly into the circuit and when the current to be measured is greater than the capacity of the instrument, and it is also used in connection with relays.

The series transformer was first considered and used in connection with street lighting systems, but was an entire failure.

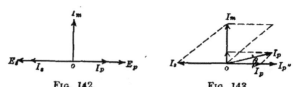

FIG. 142.   FIG. 143.

FIGS. 142 and 143.—Fundamental series transformer vector relations.

For almost 20 years it kept in the background until it came into commercial use in connection with measuring instruments.

At this time its accuracy was equal to any other instrument on the market and as the accuracy of instruments improved, the demand for a more accurate device increased. Also with the introduction of higher voltages better insulation was required, insulation that would not only protect the instruments but also the operator. As time went on it was found essential to use a magnetic circuit without joints in order to keep the magnetizing current within reasonable limits compared with the accuracy of transformation. To accomplish this a core was made of sheet-iron rings the primary and secondary windings being placed on this core. For high voltages, insulation became a serious problem. The number of turns of insulation had to be increased. This meant a considerable waste of time and money, and even after they were once in place it was difficult to take care of the heat from the windings. To overcome the necessity of using an extensive number of turns of insulation a new type was developed in which the low voltage secondary was wound on the core, while the insulating material consisted of telescoping tubes held apart by suitable spacing strips.

Up to recent years it was considered quite sufficient to insulate the series transformer with the same margin of safety as that allowed in constant-voltage transformers; that is, double the line voltage. For several reasons it was deemed advisable to require an insulation test of three times that of the operating circuit. It is evident that as the secondary is usually grounded it is liable to receive and provide a path for a lightning discharge; moreover, it furnishes current to instruments, and protecting devices, and might be a source of danger to the attendant.

The transformer consists of an iron magnetic circuit interlinked with two electric circuits. The primary is connected in series with the line, the current of which is to be measured, and the secondary is connected to instruments, etc. It is evident that the meter readings will go up and down with the primary current; though the ratio of the instrument to the primary current may not be the same at all times, any one value of the current will always give the same reading. In well designed transformers the ratio of primary to secondary current is nearly constant for all loads within the designed limits.

In the case of a series transformer with its primary connected to the line and its secondary on open circuit, the primary current will set up a magnetic field in the iron of the transformer, which will cause a drop in voltage across the primary. The same magnetic flux will also cut the secondary and generate in its winding an e.m.f. the value of which is equal to the voltage drop across the primary multiplied by the ratio of the secondary to primary turns. When the secondary is open-circuited the iron of the transformer is worked at a high degree of saturation, which produces an abnormally large secondary voltage. This condition gives rise to serious heating of the transformer as well as great strains upon the insulation.

If the secondary circuit be closed through a resistance there will be a secondary current, which allows a larger resultant flux in the core the less the value of the current, which flux generates the secondary e.m.f. An increase in the secondary resistance does not mean a proportionate decrease in the secondary current, it only means such a decrease in the current as would increase the resulting magnetic flux and secondary e.m.f. sufficiently to maintain the current through the increased resistance. Under ordinary conditions the resistance in the secondary circuit is low, so that the secondary e. m. f. is low and also the resultant magnetic flux.

If the secondary be short-circuited so that there is no magnetic leakage between the windings, and current put on the line, a magnetic flux will be set up in the primary. This flux produces an e.m.f. in the secondary which sets up a current opposed to that in the primary. The result is that the flux threading the windings will be reduced to a value which will produce a sufficient

Fig. 144.—Direction of three-phase currents—chosen arbitrarily for convenience.

voltage to establish current through the secondary resistance. Thus the magnetomotive force of the primary current is less than that of the secondary current, by an amount such that the flux produced thereby generates the voltage required to send the secondary current through the resistance of the secondary circuit, the vector sum of the secondary current and the magnetizing current being equal to the primary current.

When the secondary resistance is increased, there will be a decrease in the secondary current which allows a larger resultant flux, which in turn decreases the secondary e.m.f., which increases the secondary current. When a stable condition is reached there is a greater secondary e.m f. and a less secondary current. In order to determine the characteristics of a series transformer it is in general necessary to know the resistances and reactances of the primary and secondary windings of the transformer and of the external secondary circuit, and the amount and power-factor of the exciting current at the various operating flux densities in the transformer. If no magnetizing current were required, the secondary ampere-turns would be in

SERIES TRANSFORMERS

approximate equilibrium with the primary ampere-turns and consequently the ratio of the primary to the secondary current would be the inverse of the number of turns. In order to attain this ratio as nearly as possible the iron is worked considerably below the "knee" of the B-H curve, so that very little magnetizing force is required.

The series transformer is worked at about one-tenth the magnetic density of the shunt transformer, due to the fact that a large cross-section of iron is used. It is readily understood that a series transformer differs very much mechanically and electrically from a shunt transformer; the latter maintaining a practically constant voltage on the secondary irrespective of the load, while the former must change its secondary voltage in order to change its secondary current.

The fundational expression for the current of a series transformer is the same as for a constant potential transformer, or

$$\frac{I\,p}{I\,s} = \frac{N\,s}{N\,p} = k \text{ or } N\,s = K\,N\,p \qquad (22)$$

where

$I\,p$ = primary current. $I\,s$ = secondary current.
$N\,p$ = primary turns of wire in series.
$N\,s$ = secondary turns in series.
K = constant = ratio of transformation.

If the ratio of transformation is known and primary turns fixed, the secondary turns are equal to primary turns multiplied by the ratio of transformation.

Assuming a one to one ratio and a non-inductive secondary load, the diagram shown in Fig. 142 is obtained. The primary current and e.m.f. are equal and opposite in phase to the secondary current and e.m.f. and as the load is non-inductive the primary current is in time phase with the primary voltage.

where

$O\,I\,p$ = primary current. $O\,E\,p$ = primary e.m.f.
$O\,I\,s$ = secondary current. $O\,E\,s$ = secondary e.m.f.
  and $O\,I\,m$ = magnetizing current.

In actual operation the secondary current is in time phase with the secondary e.m.f. if the secondary load is non-inductive, but the primary current lags behind the primary e.m.f., as shown in Fig. 143 by the angle ϕ.

The secondary ampere-turns $O\,I\,s$, and the primary turns

STATIONARY TRANSFORMERS

$O\,I\,p$ being equal to the line current multiplied by the primary turns is made up of two components $I\,p$ and $I\,p'-O\,I\,p'$. $I\,p'$ being that part which supplies the core loss, and $O\,I\,p'$ that part which is equal and opposite to the secondary ampere-turns. The iron loss and "wattless" components ase $I\,p'\,I\,p''$ and $I\,p$.

In a well designed series transformer, the secondary ampere-turns for non-inductive secondary load can be taken as equal to $O\,I\,p-I\,p'\,I\,p''$, there fore secondary amperes is

$$I\,s = \frac{O\,I\,p - I\,p'\,I\,p''}{N\,s} \qquad (23)$$

To determine this, it is merely necessary to find the ampere-turns $O\,I\,p$ supplying the iron loss, subtract this value from the primary ampere-turns $I\,p$ and $I\,p'$, and divide by the secondary turns $I\,s$.

To determine the iron loss it is necessary to know the loss in watts per unit weight of iron at a given frequency and varying induction. Let P represent the watts loss at a given frequency and induction, I the iron loss current, $N\,s$ secondary turns, then

$$\text{Iron loss current} = \frac{P}{E} \qquad (24)$$

and

$$\text{ampere-turns (iron loss)} = I\,N\,s = \frac{P\,N\,s}{E} \qquad (25)$$

$$\text{secondary ampere-turns} = I\,p\,N\,p - \frac{P\,N\,s}{E} \qquad (26)$$

and

$$\text{secondary amperes} = I\,p\frac{N\,p}{N\,s} - \frac{P}{E} \qquad (27)$$

If the ratio of transformation is made equal to $\frac{I\,s}{I\,p}$ the secondary current will be less than the desired value by the amount $\frac{P}{E}$ amperes. The error in the transformation is

$$\text{per cent. error of } \frac{I\,s}{I\,p} = \frac{P \times 100}{I\,s \times E} \qquad (28)$$

To compensate for this error the secondary turns must be slightly diminished from the ratio $\frac{I\,p}{I\,s}\,N\,s$ so that the secondary current will equal $I\,s$.

Assume that the secondary is on short-circuit, and for convenience, that there is no magnetic leakage between primary and secondary. At the moment the current is started, flux is set up in the primary winding. This flux produces an e.m.f. in the secondary which sets up a current opposed to that in the primary. The result is that the flux threading the two windings will be reduced to a value producing only sufficient voltage to cause current through the secondary resistance, thus restoring approximate equilibrium between the primary and secondary currents. Thus the magnetomotive force of

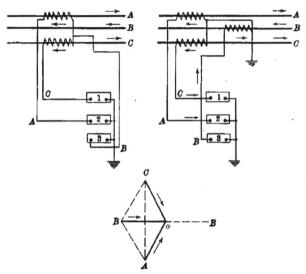

FIG. 145.—Method of obtaining equal three-phase current readings by the use of two- or three-series transformers.

the primary current is less than that of the secondary current by an amount such that the flux produced generates the voltage required to send the secondary current through the resistance the vector sum of this current and the magnetizing current being equal to the primary current.

A series transformer should maintain a practically constant ratio between its primary and secondary through its full range of load. Such a condition can only be approached but not absolutely reached, since the magnetizing current becomes a

192 STATIONARY TRANSFORMERS

formidable factor in preventing a constant ratio. A minimum magnetizing current is accomplished in commercial transformers by having an abundance of iron in the core, thus working the iron at a very low magnetic density, permitting the current ratio of the primary to the secondary to vary approximately in inverse ratio to the number of turns. As iron is worked considerably below the "knee" of the B-H curve, a good range of load is allowed for ammeters, wattmeters and relays to be operated on the secondary.

In either a delta- or star-connected three-phase system the currents in the three leads are displaced 120 degrees from each other and one of the leads may be considered a common return for the other two. Assuming the instantaneous direction of current in leads A and C to follow the arrows shown in Fig. 144, then the direction of the current in B will follow the arrow in the opposite direction. With a delta-connected system the current in lead A is the resultant of that in the two phases BA and CA; the current in lead B is the resultant of that in the two phases AB and CB; and the current in C is the resultant of that in AC and BC. With the star connection, the direction of the two currents A and C are in opposite directions to the neutral point, and the current B toward the neutral point O.

If all the secondaries of series transformer are not arranged in the same direction, it will be found that the phase relation between the three phases will be either 60 or 120 degrees. In either a delta- or star-connected group of transformers it becomes necessary to reverse the secondary leads with respect to the others when a change of phase relation is desired. A connection comprising two series transformers is shown in Fig. 145 where A and C are connected to two ammeters, and the opposite end of secondary winding connected to a common wire at O. In the vector diagram it is shown that OA is equal to the current in lead A both in magnitude and direction and OC the current in lead C which has a phase relation of 120 degrees. The resultant current that is in the common wire O is equal in magnitude and direction to OB'. The ammeter (2) indicates the current in lead B, and its reading is equal to the value shown in Fig. 145.

Assume that one of the transformers is reversed as at C, Fig. 146. Referring to the vector diagram, OA represents the current in lead A both in value and direction, OC the current in lead C, and the resultant current $A\,C$ represents in value and direction the current through the ammeter (2). The resultant current $A\,C$ is displaced from that of lead B by 90 degrees, and is found to be $\sqrt{3}$ as great as $A\,O$, $O\,B$ or $O\,C$. Thus, if the two transformers $O\,A$ and $O\,C$ are wound to give 5 amperes on their secondaries with full load current in the primary, the current across $A\,C$ will be $5\times\sqrt{3} = 8.66$ amperes. So that in connecting series transformers and it is found that one of the phase currents bear $\sqrt{3}$ relation, it is simply necessary to reverse one of the transformer secondary leads.

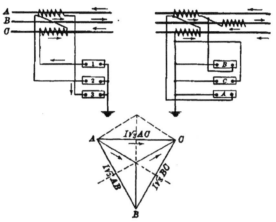

Fig. 146.—Another method of using two- and three-series transformers with three ammeters.

The volt-amperes on each transformer of a two-transformer connection like Fig. 147 are equal to $\sqrt{3}$ times the volt-amperes of load $A\,B\,C$; but the phase angle between voltage and current is charged 30° in the lagging direction on transformer A, and in the leading direction on transformer B. For power-factor of the secondary loads A, B and C varying from 100 per cent. to 0 per cent., the power-factor of the equivalent load on transformer A will vary from $\sqrt{3}$ leading to $1/2$ lagging; while on transformer B it will vary from $\sqrt{2}$ lagging to a negative $1/2$.

Where (a) is for equal loads on each phase and equal power-factors.

(b) Equal loads and non-inductive.

(c) Equal loads (loads A and B non-inductive) and C 50 per cent. power-factor.

(d) Equal loads (loads A and B non-inductive) C 10 per cent. power-factor.

A tendency with lagging power-factor in load. C is to increase the equivalent load on the transformer A which is connected to the leading phase, and to diminish the equivalent load on transformer B which is connected in the lagging phase. Low power-factor on both A and B combined with high power-factor in C produces similar results.

There are a number of different ways of connecting two or

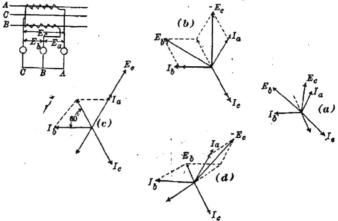

Fig. 147.—Varying phase relations due to varying loads and power factors through the series transformers.

more series transformers to a polyphase system. One or more may be used in connection with alternating-current relays for operating circuits for overload, reverse power, reverse phase, and low voltage. Some of the connections for this purpose are shown in Figs. 148 to 149.

One series transformer is sufficient for opening the circuit of a single-phase system, and at the same time is used in connection with an ammeter and wattmeter as shown in Fig. 148. For three-phase working, one, two or three series transformers may be used for relays, ammeters and wattmeters. It is often recommended that three series transformers should be used for three-phase systems, but in the majority of cases two are sufficient to give satisfactory results. In Fig. 149 one is shown connected

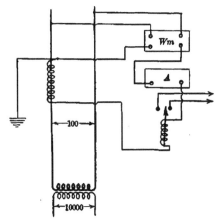

Fig. 148.—Simple series transformer connection.

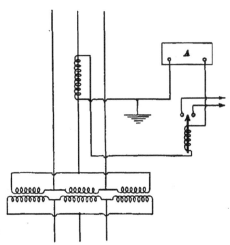

Fig. 149.—Method of connecting a series transformer, ammeter, and relay on a delta connected three-phase system.

to one leg of a three-phase delta system. Its secondary circuit is connected through a relay and an ammeter.

In connection with overload relays one series transformer may be used for operating a three-phase system, and when operating three-phase wattmeters, two are all that is required. The connection shown in Fig. 150 will be found to give good results, as each transformer has its own tripping arrangement.

Three transformers are quite common operating together on a three-phase, star-connected system, neutral point grounded or ungrounded.

If all the secondary windings are not arranged in the same direction the phase relations between one outside wire and the

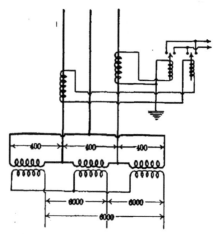

Fig. 150.—Method of connecting two series transformers and relays to a three-phase system.

middle wire, and the middle and the other outside wire will be 60 degrees instead of 120 degrees. In order to obtain a phase relation of 120 degrees between each winding, one of the secondary windings must be reversed.

Fig. 151 represents three series transformers with all the secondary windings connected in one direction. It makes no difference whether the method of connection be delta or star, it becomes necessary to reverse one transformer with respect to the others when 120 or 60 degrees displacement is required.

A connection very much used where one relay is required, is shown in Fig. 152, in which the series transformers have

SERIES TRANSFORMERS

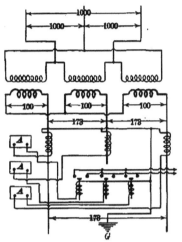

Fig. 151.—Method of connecting three series transformers, three an and three relays on a three-phase star connected system.

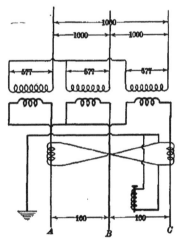

Fig. 152.—Three-phase star arrangement showing two series trans connected in opposition.

their opposite terminals connected. The secondary phase relations tend to operate in parallel so that when a current exists in the primary of one transformer a current will also exist in the secondary and relay, but will not be great enough to operate the trip coil. If a short-circuit should occur on any one phase of the two outside wires $A\,C$, the secondary will become overloaded and its voltage will rise to a value above that of the secondary of the other transformer; this will tend to reverse the current in the latter transformer, which in turn will allow the

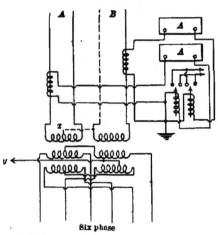

Fig. 153.—Method of connecting two series transformers with instruments and relays to a two- or three-phase system inducing six-phase secondary currents.

primary flux to raise the voltage to the value of the former transformer; this voltage will cause the additional current to overload and operate the relay. This current value will not be twice the current through the two transformers, but will be the algebraic sum of the currents at 120 degrees apart, or $\sqrt{3}$ times the current in each leg. For example: If two series transformers are wound for 5 amperes on their secondaries with normal current through their primaries, the algebraic sum of the two currents is

$$\sqrt{3} \times 5 = 8.66 \text{ amperes.}$$

Fig. 153 shows a two-phase arrangement of connecting two

SERIES TRANSFORMERS

series transformers for working instruments and relays. It will be noticed in all the connections shown that the secondaries of all transformers are grounded on one side.

In Fig. 153 the system is so arranged that two-phase or three-phase currents will give a six-phase secondary, depending upon the connection made at point x. The series transformer connections are so arranged that the instruments will work satisfactorily with any of the two independent phase currents.

The Y represents the neutral point of the secondary power

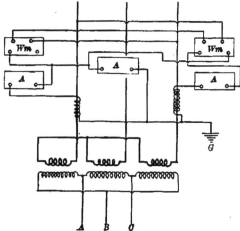

FIG. 154.—Three-phase star arrangement showing two series transformers, two wattmeters, and three ammeters.

transformers, and may be used as a neutral wire in connection with a direct-current system of supply.

Another interesting connection is shown in Fig. 154, in which currents are measured in the three phases by the use of two series transformers. The geometrical sum of the currents in the primaries where the two series transformers are installed, is measured by the ammeter shown connected to the grounded side of the transformers. The value obtained is that of the current through the middle wire.

The potential sides of the two wattmeters may be connected to the secondary leads of two-shunt transformers; in the figure they are shown connected directly to the mains.

In general, single and polyphase combinations of series transformer connections are covered by the use of one to four transformers. In three-phase work either the star, delta, open-delta, reversed open-delta and "Z" connections might be applied.

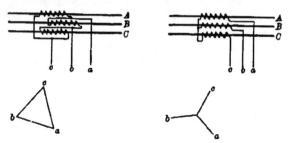

Fig. 155.—Three transformer delta and star connections.

(See figs. 155, 156, 157.) The connection shown in Fig. 155 is not unlike the ordinary constant potential transformer delta. In this connection the currents in the secondary leads A and B are the same, but the current through C is $\sqrt{3}$. For relays it is thought to be much better than the delta connection.

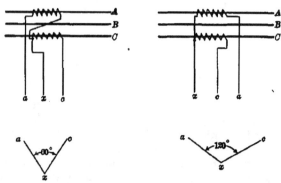

Fig. 156—Two transformer "V" and inverted "V" connections.

It is also possible to measure three-phase currents with two series transformers and only one ammeter. The arrangement is shown in Fig. 158. To read the current through the transformer on the left, the two switches, a and b, are closed. To read the current in the middle line, b and c are closed; the current through the transformer on the right is measured by closing the two switches, c and d.

When measurements are not being taken it is necessary that the switches, a and d, should be closed; as the iron of the two transformers is worked at a high degree of saturation, which produces an abnormally large secondary voltage, giving rise to a serious heating of the transformer.

Since series transformers are connected directly in series with the line, if not properly installed, they will offer a convenient path for the escape of high frequency charges which may occur on the line, and which in discharging, not only burn out the transformer, but are likely to form an arc and probably cause a fire, or loss of life.

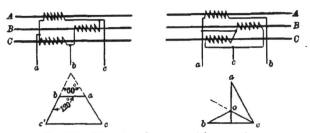

FIG. 157.—Three transformer special connections.

The process of drying out air-insulated series transformers is accomplished by simply passing normal current through the winding until the transformer is thoroughly warmed. This may be done by short-circuiting the secondary through an ammeter and sending enough alternating current through the primary to give normal current on the secondary, the primary current being obtained from a low voltage source. If not convenient to obtain low voltage alternating current the same result may be accomplished by passing normal direct current through the primary long enough to thoroughly warm the transformer.

In case of oil-cooled transformers, the winding should be dried out by this same process before the transformer is filled with oil. In doing this the temperature of the coils should not be allowed to exceed 65° C., which may mean the use of a current much less than normal, owing to the fact that there is no oil in the transformer.

All secondaries and casings of transformers should be grounded, and likewise the instruments to which they are connected. This serves the double purpose of protecting the switchboard attendant and freeing the instruments from the effects of electrostatic charges which might otherwise collect on the cases and cause errors.

If for any reason it becomes necessary to remove an instrument or any current carrying device from the secondary circuit of a series transformer, the secondary should be short-circuited

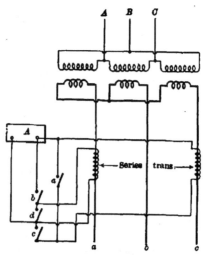

FIG. 158.—Method of connecting two series transformers and one ammeter to a three-phase system, to measure the current in any lead.

by a wire or some other means. Series transformers should be considered as a part of the line circuit. When it becomes necessary to change secondary connections, the ground wire should be inspected to see that it is in good condition. The operator should also stand in a dry board or other insulating material.

By reason of phase displacement in series transformers, wattmeters, when used with series transformers, will have certain errors due to such displacement. The following table applies to certain types of series transformers, the actual percentage of error varying slightly in different manufacturers for the same rating in capacity.

When a wattmeter is used in a circuit for 100 per cent. *power-factor*, the approximate errors will be:

TABLE VI

Transformer capacity (watts)	1-5 Watt load or 1-5 current load	1-2 Watt or current load	Full watt or current load
(a) 2	−2.0%	−2.0%	−2.0%
(b) 10	+0.5%	−0.5%	−1.5%
(c) 30	+0.5%	−0.5%	−1.5%
(d) 30	+0.5%	−0.5%	−1.5%

The current ratios remain true and cause no error when used with ammeters between the limits of one-tenth load and 50 per cent. overload.

When potential transformers are used as well, the tendency of the two (series and potential transformers) is to neutralize the error.

When a wattmeter is used in a circuit of 50 per cent. *power-factor*, the approximate errors may be:

TABLE VII

Transformer capacity (watts)	1-2 current or watt load	1-5 current or watt load	Full current or watt load	Double current or watt load
(a) 2	+4.0%	+12.0%	+2.0%	+2.0%
(b) 10	+4.0%	+ 7.0%	+1.5%	−1.5%
(c) 30	+2.0%	+ 5.0%	0.0	−0.5%
(d) 30	+2.0%	+ 5.0%	0.0	−0.5%

If the wattmeter is calibrated with the series transformer at 100 per cent. *power-factor*, the error at 50 per cent. *power-factor* may become:

TABLE VIII

Transformer capacity (watts)	1-2 current or watt load	1-5 current or watt load	Full current or watt load	Double current or watt load
(a) 2	+6.0%	+14.0%	+4.0%	+4.0%
(b) 10	+4.5%	+ 6.5%	+3.0%	0.0
(c) 30	+2.5%	+ 4.5%	+1.5%	+1.0%
(d) 30	+2.5%	+ 4.5%	+1.5%	+1.0%

At *power-factors* of less than 50 per cent. the errors will greatly increase.

It will be seen that if the wattmeter is calibrated with the series transformer, greater accuracy may be obtained when used in circuit of some power-factor, but that the erroe at lower power-factor becomes greater.

The plus sign (+) means that the wattmeter will indicate more than the true power by the percentage shown. The minus sign (−) means that the wattmeter will indicate error in the opposite direction.

Excepting (d) which indicates a wound primary type transformer, (a), (b) and (c) are for transformers of the open type intended to slip over bus-bar or switch-stud as well as the wound primary type transformer (5 amp. secondary).

It is sometimes desired to use for meter test and other purposes, inverted series transformers; that is, "step-up current transformer." Fig. 159 shows an arrangement using two series transformers of any desired ratio but equal; No. 2 transformer is inverted, and, depending on the ratio of transformation, may be employed to step up the current to any desired value. Assuming that both transformers are for 40 to 1 ratio (200–5 amp.), it is evident, from the diagram that the standard testing instrument in No. 1 transformer circuit has a 5-ampere current coil and the service meter 200 amperes in its current coil when a 5-ampere load is flowing through the circuit load *b*.

For some time past, the series transformer has been used for low-voltage series street lighting in connection with series arc lighting systems.

For the purpose of lighting, its primary winding is connected in series with the series arc lighting system, so that under all conditions of load on the secondary the primary winding carries the full current of the arc circuit (see Chapter XI) which is maintained at its normal value by the constant current transformer. Its connections are shown in Fig. 160.

For satisfactory operation of the series incandescent lamp, it is desirable to obtain as near constant current as possible in the secondary winding. Of course, to obtain constant-current regulation under abnormal conditions of load is impossible, but it has been found in practice that with open circuit voltage on the secondary not exceeding 150 per cent. of the full-load voltage, and the current at 100-75 and 50 per cent. load not varying more than 2 per cent., the regulation of the transformer and lamps on the secondary is satisfactory.

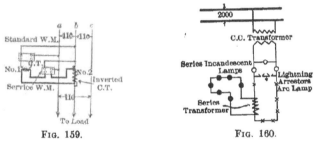

Fig. 159. Fig. 160.

Fig. 159.—Series transformer used to step-up current, as for instance, with a primary load of 5-40 volt lamps, 700 amperes are obtained.

Fig. 160.—Method of using series transformer or series arc lighting circuits for low-voltage series incandescent lighting.

Certain classes of lighting require lower potential than that obtained from series arc circuits, and to provide for this, light and power companies are often compelled to run multiple circuits from the central station or substation at a considerable expense. By using a series transformer on the series arc lighting circuit, a low-voltage circuit may be run when required, thus obviating a large item of expense and providing a very flexible system of distribution.

The ratio of transformation of series transformers used for lighting purposes has generally been 1-1, bur there is no difficulty in winding either primary or secondary for any reasonable current.

. The core of this transformer is of the shell type, built up of circular punchings with two symmetrical pieces in each layer. On the center leg or tongue of the core are assembled the form-wound coils. The primary coil fits simply over the secondarp coil, but is so insulated that it will withstand a break-down test of 20,000 volts to the secondary coil and also to the core. Its appearance is not unlike the well-known telephone line insulating transformer which is used in connection with long-distance high-voltage transmission work.

In this series transformer there exists a drooping voltage characteristic in the secondary. Its purpose is to limit the open-circuit voltage on the secondary. It has been obtained by so proportioning the magnetic circuit that the section is contracted in several parts to permit saturation of the iron with no current in the secondary winding.

There exist quite a number of meter connections which may be used on power transformer systems. Some of these are given below: series and potential transformers being used in each case.

Fig. 161(A) shows the simplest three-wire three-phase meter arrangement, in which a single-phase wattmeter has its potential coil connected to the secondary of a potential transformer, and its current coil to the secondary of a series transformer. If the power transform should have an accessible neutral point, at the location x, the potential transformer may be connected according to the dotted line.

Fig. 161(B) shows the next simplest three-wire three-phase wattmeter arrangement, in which two single-phase wattmeters are involved. This method of measuring electrical energy in three-phase circuits for any condition of unbalanced voltage or current is the correct one.

Fig 161(C) connection is for a three-wire three-phase system. The wattmeter has its current coil connected to one phase through a series transformer (c. t.), and its potential coil to a "Y" impedance, each branch of which has equal reactance and resistance such as to give the proper voltage to the potential coil of the meter.

Fig 161(D) connection is not unlike (B). The polyphase or two-element wattmeter is employed here, and should always be used in preference to two single-phase meters. It has a slightly better accuracy if we consider that the record of the meter as a whole is obtained without the possibility of error that comes from

SERIES TRANSFORMERS

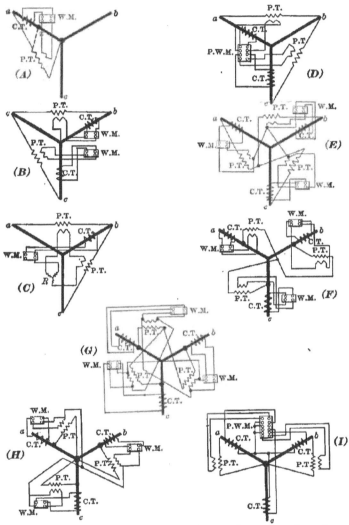

Fig. 161.—Other important uses for series transformers on three-phase systems.

reading the two single-phase meters separately and adding their sum total together afterward.

Fig 161(E) and (F) show two arrangements of the three single-phase wattmeter method for a three-phase three-wire system. The potential transformers are connected in star-star. This method is not recommended because of unequal loading which might occur on the secondary side. With unbzlanced load on the secondary, the voltages might become very unequal, and with an accidental short-circuit on one transformer its primary impedence may be so reduced that the remaining two transformers would be subjected to almost $\sqrt{3}$ times their normal voltage.

Fig 161(G) shows a three-wattmeter method in which three series transformers, and three potential transformers are used, the latter being connected in star-delta. This method should always be used in preference to methods (E) and (F) for the reason the balanced voltages on the primary are established by exchange of current through the secondary circuits of the three transformers in case the secondary loads are unbalanced. Abnormal voltages are prevented in the individual transformers and meters.

Fig. 161(H) shows a four-wire three-phase, three-wattmeter method. The potential transformers are connected in star-star, but the objection referred to in (E) and (F) is practically eliminated. The arrangement is the same as three independent single-phase systems with one conductor of each phase combined into a common return. This method of measuring energy is considered to be superior to the three-wattmeter method shown in (E), (F), and (G), for the reason that no inter-connection of potential transformers is required and because the wattmeters cannot be subjected to variations of voltage greater than those properly belonging to the circuit.

Fig 161(I) shows a four-wire, three-phase, polyphase wattmeter method. This method is preferable to method (H) and offers greater simplicity and convenience.

CHAPTER XIII

REGULATORS AND COMPENSATORS

Potential Feeder Regulators.—Almost all regulators are of the transformer type, with their primary windings connected across the lines and their secondary windings connected in series with the circuit the voltage of which is to be controlled.

A type of single-phase feeder regulator is shown in Fig. 162. It consists of a laminated iron ring with four deep slots on its inner surface, in which the primary and secondary windings are placed. The laminated core is mounted on a spindle and

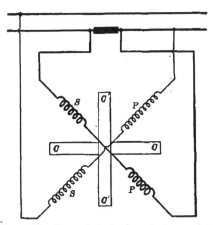

Fig. 162.—Type of single-phase feeder regulator.

so arranged that it can be turned to any desired position by means of a hand wheel. In the position indicated by $C\ C$, the core carries the magnetic flux due to the primary winding, P, in one direction through the secondary winding, S; and in the position indicated by $C'\ C'$ the core carries the magnetic flux due to the primary winding, P, in the other direction through the secondary winding, S. That is, when the core is in the position $C\ C$ the generated voltage in the secondary winding has its highest value. When the core is midway between $C\,C$ and $C'\,C'$ the generated voltage in the secondary winding is zero, and the feeder voltage is not affected. When the core is in the position, $C'\,C'$, the generated voltage in the secondary winding is again at its greatest value, but in such a direction as to oppose the generator voltage.

On account of the air-gap between the primary and secondary windings, inductive reactance is introduced in the line which requires compensating.

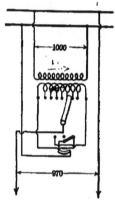

Fig. 163.—Type of Stillwell regulator.

The Stillwell regulator is another type of transformer for raising and lowering the voltage of feeder circuits. It consists of a primary winding which is connected across the feeder circuit, and a secondary winding in series with the circuit the voltage of which is to be varied. By means of a switch arm, more or less of the secondary winding may be introduced into the circuit, thus "boosting" by a corresponding amount the voltage of the generator. A reverse switch is provided to which the primary winding is connected, so that the voltage may be added by moving the switch arm to the right, and subtracted by moving the switch arm to the left.

A regulator built along the lines mentioned above, with an arrangement for connecting the various sections of the secondary winding to a dial switch and reversing switch, is shown in Fig. 163. The feeder potential can be controlled in the following manner: Starting with the regulator in position of maximum boost, that is, with the dial switch turned to the extreme left as far as it will go, a continuous right-hand movement of the dial switch for two complete revolutions is obtained. During the first revolution the switch cut outs, step by step, the ten sections of the secondary winding. When the first revolution has been completed, the voltage on the feeder is the same as that of the generator, no secondary winding being included. A further movement of the switch in the same direction automatically throws a reversing switch; and continuing the movement of the dial switch, still in the same direction, the secondary windings are again switched in, step by step, this time with reversed polarity; so that when the second revolution is complete the whole secondary winding is again included in the feeder, but now opposing the voltage of the generator. Thus by one complete movement of the switch, covering two revolutions in one direction the complete range between the maximum boost and maximum depression of the feeder voltage is covered.

In incandescent lighting service a potential regulator is particularly valuable. Within the ordinary limits of commercial practice the candle-power of an incandescent lamp will vary approximately 5 per cent. for every 1 per cent. variation in the

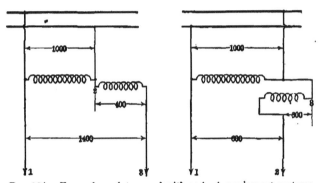

Fig. 164.—Type of regulator used with series incandescent systems.

voltage at the terminals. That is to say, if a 16-c-p. 100-volt lamp be burned at 106 volts it will give about 21 c-p., or at 94 volts about 11 c-p. This fact shows at once the urgent necessity for keeping the voltage of an incandescent system adjusted within inperceptible degrees. A method that has given somewhat satisfactory service with series incandescent systems is shown in Fig. 164. It consists in a primary winding which is connected across the main lines. Attached to one end of the primary winding is a secondary winding. The secondary voltage, 1, 3, is greater than the primary, 1, 2, by the voltage of the winding, 2, 3. The voltage, 2, 3, is thus added to the primary to form the secondary voltage of the circuit. By reversing the connections of the windings, 2, 3, it may be made to subtract its voltage from the primary, 1, 2, in which case the secondary voltage, 1, 3, becomes less than the initial primary voltage; see

Fig. 165. Further, by bringing a number of leads from parts of the winding, 2, 3, the secondary voltage, 1, 3, may be increased or decreased step by step as the different leads of 2, 3, are connected to the secondary circuit.

In Fig. 165 is shown the connections of this form of regulator, or compensator. 1, 2, represents the primary winding connected across the circuit. From a portion of the secondary winding, 3, 4, taps are brought to the contact blocks shown in the diagram. The two arms, 7, 8, connect these contact blocks to two sliding contacts, 5, 6. The arms, 7, 8, may be operated by a handwheel, one in direct contact and the other through a gearing, so that a rotation of the handwheel turns one arm clockwise and the other counter-clockwise.

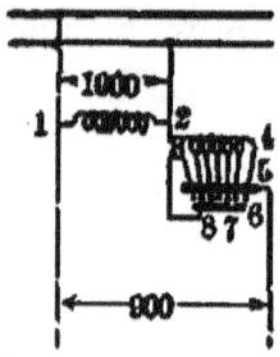

FIG. 165.—Type of series incandescent regulator that reduces the initial primary voltage.

The secondary voltage of the circuit is that of the primary, increased or decreased by the voltage between the arms, 7, 8. In the neutral position, both arms rest on one central contact block, and the difference of potential between them is zero. In order to decrease the voltage at the lamps, the handwheel is turned to the right, and the voltage decreased step by step, until the final position is reached with each arm on an extreme contact block. To increase the voltage at the lamps, the handwheel is turned to the left, the two arms being gradually separated on the contact blocks; and the difference of potential between the arms is effected step by step until in the final position, each arm rests on an extreme contact, and the secondary winding is connected into the circuit and its total voltage thus added to the initial voltage of the system. This type of regulator, as will be seen, is in the order of an ordinary auto-transformer with regulating taps arranged on its secondary winding.

The induction type regulator differs from the transformer type in that all the primary and secondary windings are constantly in use. There are types that vary the secondary voltage either by moving part of the iron core or one of the windings, or one of the windings (primary or secondary) and part of the iron core; the whole or part of the magnetic flux generated by the primary threads the secondary according to the position of the moving part.

This type of regulator is either self-cooled oil immersed, oil-immersed water-cooled, or forced air-cooled depending on the capacity.

Single-phase regulators have only one excitation winding, the magnetizing flux is an alternating one and its direction is always parallel to that diameter of the movable core which passes through the center of the exciting coil, but its direction may be varied with respect to the stationary core, and, consequently, with the respect to the stationary or series winding.

With the armature in such a relation to the field that the primary winding induces a flux opposed to that induced by the secondary, the voltage induced by the primary in the secondary is added directly to the line voltage, but is subtracted when the direction of the flux is the same, the complete range being obtained by rotating the armature through an angle of 180 degrees. As the core is rotated gradually, the relative direction of the primary flux, and, consequently, the amount forced through the secondary coils, is similarly varied and produces a gradually varying voltage in the secondary from the maximum positive, through zero, to the maximum negative value. The induced voltage, is, however, added directly to, or subtracted directly from the line voltage.

The primary or rotating core contains two windings; the active windings connected across the line, and a second winding short-circuited on itself and arranged at right angles to the active winding. The object of this short-circuited winding is to decrease the reactance of the regulator, and its operation is as follows: As the primary and the short-circuited windings are both on the movable core and permanently fixed at right angles to each other, the flux generated by the primary passes on either side of the short-circuited coil, and is, therefore, not affected by it in any way whatever; for as long as no flux passes through this coil there is no current in it. This condition is, however, only true when the armature is in the maximum boost or maximum lower position with current in the series winding, and in any position of the armature with no current in the secondary.

With the armature in the neutral or no boost or lower position the flux generated by the current in the secondary passes equally on either side of the primary coils, which cannot, therefore, neutralize the flux generated by the secondary.

If the primary core were not provided with a short-circuited winding, and rotated from maximum position so as to reduce the primary flux passing through the secondary, and if the line current remained constant, a gradually increasing voltage would be required to force the current through the series windings, and a correspondingly increasing flux would have to be generated. This voltage would become a maximum with the armature in the neutral position, due to the fact that in this position the primary windings are at right angles to the series windings and therefore entirely out of inductive relation to them. The current in the secondary, therefore, would act as a magnetizing current, and a considerable part of the line voltage would have to be used to force the current through these coils. The voltage so absorbed would be at right angles to the line voltage, and the result would be a poor power factor on the feeder.

The short-circuited coil on the armature which is in a direct inductive relation to the series coils when the armature is in the neutral position, acts as a short-circuit on the secondary winding,

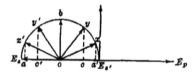

Fig. 166.—Phase positions of primary and secondary voltages of a single-phase induction regulator.

and thereby reduces the voltage necessary to force full load current through this winding to only a trifle more than that represented by the resistance drop across the secondary and short-circuited windings. This short-circuiting of the secondary is gradual, from zero to the maximum boosting position of the regulator to the maximum short-circuiting in the neutral position, so that by the combined effect of the primary and short-circuited coils the reactance of the secondary is kept within reasonable limits.

The operation of the short-circuited coil does not increase the losses in the regulator, but rather tends to keep them constant for a given secondary circuit. In rotating the armature from either maximum to the neutral position, the current in the primary diminishes as the current in the short-circuited coil increases, so that the total ampere-turns of the primary plus the ampere-turns of the short-circuited winding are always approximately equal to the ampere-turns of the secondary.

In Fig. 166 there is shown graphically the values and time-phase position of primary and secondary voltages of a single-phase induction regulator.

When the mechanical position in electrical degrees of the moving part is shifted to $O\ Y$ on $O\ X$ following the curve of the semi-circle in the position of negative boost, or $O\ Y'$ and $O\ X'$ in the position of positive boost, the secondary voltage can be considered to have the values $O\ C$ and $O\ A$ respectively, and $O\ C'$ and $O\ A'$ respectively.

When the mechanical part occupies a mechanical position of 90 electrical degrees from the position $O\ E\ s'$ and $O\ E\ s$ the value $O\ B$ of the secondary voltage is zero, because the flux due to the primary exciting current passes through the secondary core parallel to the secondary windings. The resultant voltage is equal to the primary voltage.

The kilowatt capacity of any regulator is equal to the normal line current to be regulated times the maximum boost of the regulator, and as the lower is always equal to the boost, the total range is equal to twice the kilowatt capacity or 100 per cent. with a 1 to 2 ratio.

Kilowatt capacity of a single-phase regulator is maximum boost or lower times the current, or the maximum boost or lower might be expressed in terms of kilowatts divided by the secondary current. For a two-phase type it is one-half this amount. In a three-phase type the boost or lower across the lines is equal to the regulator capacity in kilowatts multiplied by $\sqrt{3}$ and divided by the secondary current.

Regulators of the induction type should not be used for any other primary voltage or frequency differing more than 10 per cent. from that for which they are designed, because an increase in voltage or a decrease in frequency increases the magnetizing current and the losses and an increase in frequency increases the impedance. The deviation of 10 per cent. allowed must not occur in both the frequency and the voltage unless these deviations tend to neutralize each other. For instance, a regulator should not be subjected to a 10 per cent. increase in voltage and the same per cent. decrease in frequency, but it will operate satisfactorily if both voltage and frequency are increased within the amount given.

Induction regulators may be operated by hand, either directly or through a sprocket wheel and chain; by a hand-controlled motor, or automatically-controlled motor. If operated by hand controlled motor, the motor may be of the alternating-current or direct-current type, but preferably of the alternating-current type and polyphase. If automatically controlled, the operating motor should preferably be of the polyphase type as the direct-current motor is not very well adapted for this purpose. When the regulator is operated by a motor, the motor should be controlled by a double-pole double-throw switch mounted on the switchboard or in any other convenient location. Closing the switch one way or the other will start the motor so as to operate the regulator to obtain a boost or lower in the line voltage as may be desired, and when the correct line voltage is obtained the regulator movement should be stopped by opening

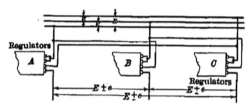

FIG. 167.—Connections of single-phase regulators operating on a three-phase system.

the switch. Generally a limit switch is provided which stops the movement of the regulator by opening the motor circuit as soon as the regulator has reached either the extreme positions depending on the direction of rotation, but which automatically closes the circuit again as soon as the regulator armature recedes from the extreme positions. The operation of each limit switch does not interfere with the movement of the regulator in the direction opposite to which it may be going.

When a single-phase regulator is used in one phase of a three-phase system, the secondary wiring is connected in series with the line and the primary between the lines, see Fig. 167. Under these conditions there is a difference in phase between the current in the two windings and the effective voltage of the secondary is therefore reduced from its normal value.

If three single-phase regulators are used, each phase can be adjusted to the range equal to the effective range of the regulator, so that the voltage between the phases is not that due to the effective voltage per regulator, but that due to the effective voltage of each regulator times $\sqrt{3}$. In this case, if 10 per cent. regulation across the phases of a three-phase three-wire system is desired, only 57.7 per cent. of the 10 per cent. regulation per line is needed; thus, the boosting or lowering need not have a greater rating than 3 to 7.5 per cent. instead of 10 per cent. which is the size necessary where one regulator is used.

In general the capacity in kw. of a three-phase regulator is rated at $E.I.\sqrt{3}$, where E is the volts boost or lower and I the amperes in feeder circuit. For six-phase rotary converter service, kw. capacity of regulator is: (double-delta connection) $=E. I. 3.46$ (29) and (diametrical connection) $= E. I. 3$ (30).

For polyphase circuits the system may be regulated by introducing the so-called "induction regulator." This form of regulator has a primary and a secondary winding. The primary winding is connected across the main line, and the secondary winding in series with the circuit. The voltage generated in each phase of the secondary winding is constant, but by varying the relative positions of the primary and secondary, the effective voltage of any phase of the secondary on its circuit is varied from maximum boosting to zero and to maximum lowering. In order to avoid the trouble of adjusting the voltages when each phase is controlled independently, polyphase regulators are arranged to change the voltage in all phases simultaneously They can be operated by handwheels or motors. When operated by hand, the movable core is rotated by means of a handwheel and shaft. When it is desired to operate the regulator from a distant point, the apparatus is fitted with a small motor which is arranged through suitable gearing to turn the movable core. The motor may be of the direct-current or induction type, and controlled at any convenient place.

The theory of this form of regulator is described graphically in Fig. 168 in which the voltage of one phase of the regulator is, $eo=$ generator voltage or the e.m.f. impressed on the primary: $ao=$ e.m.f. generated in the secondary windings, and is constant with constant generator e.m.f.: $b'a'=$ secondary e.m.f. in phase with the generator e.m.f.: $e'a'=$ line e.m.f., or resultant of the generator e.m.f. and the secondary e.m.f.

The construction of the regulator is such that the secondary voltage, oa, is made to assume any desired phase relation to the primary e.m.f., as of, ob, oc, etc.

When its phase relation is as represented by of, which is the position when the north poles and the south poles of the primary and secondary windings are opposite, the secondary voltage is in phase with the primary voltage and is added directly to that of the generator.

The regulator is then said to be in the position of maximum boost, and by rotating the armature with reference to the fields, the phase relation can be changed to any extent between this and directly opposed voltages. When the voltage of the secondary is directly opposed to that of the primary, its phase relation is as

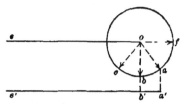

FIG. 168.—Graphical representation of an induction regulator.

represented by od in the diagram, while ob represents the shape relation of the secondary when in the neutral position.

The electrical design of the induction regulator is very similar to that of the induction motor. Its efficiency is somewhat higher than the average induction motor of the same rating. The primary winding is placed on the movable core and has either a closed delta or star connection, while the secondary or stationary winding is placed on the stationary core and is an open winding, each section or phase being connected in series with the corresponding phase of the line.

The maximum arc through which the primary moves is 60 degrees for a six-pole and 90 degrees for a four-pole. Induction potential regulators are built for single-phase, two-phase, three-phase and six-phase circuits.

Compensators are used in connection with starting alternating-current motors, and to some extent they are used in connection with voltmeters in the generating station.

Compensators for starting alternating-current motors consist of an inductive winding with taps. For polyphase work the compensator consists of one coil for each phase $a\,b\,c$, Fig. 169 with each coil placed on a separate leg of a laminated iron core. Each coil is provided with several taps, so that a number of voltages may be obtained, any one of which may be selected for permanent connection to the switch for starting the motor.

When the three-phase winding is used the three coils are connected in star, the line is connected to the three free ends of the coils, and the motor when starting is connected to the taps as represented in Fig. 169.

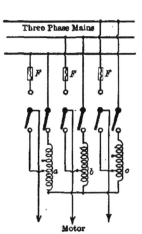

FIG. 169.—Three-phase motor compensator.

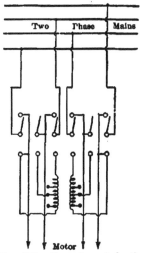

FIG. 170.—Two-phase induction compensator.

As it is difficult to predetermine the best starting voltages for each case, for motors rated at from 5 to 15 h.p., taps of 40, 60, and 80 per cent. of the line voltage are provided, according to individual requirements.

The most essential part of the compensator is an auto-transformer, the principle of which has already been explained. The switch for operating the starting compensator is immersed in oil. In starting, the switch moves from the off position to the starting position, where the lowest voltage is applied to the motor, or the position where the starting torque is the lowest that can be obtained. As soon as the motor speeds up, the switch may be thrown over to the running position; the compensator winding is then cut out and the motor is connected to the line through suitable fuses or circuit breakers. The switch is generally provided with a safety device which is used to prevent the operator from throwing the motor directly on the line, thereby causing a rush of current.

Compensators are designed to bring the motor up to speed within one minute after the switch has been thrown into the starting position. It is important that the switch be kept in the starting position until the motor has finished accelerating, to prevent an unnecessary rush of current when the switch is thrown to the running position.

In two-phase compensators the line is connected to the ends of the two coils, and the starting connections of the motor to the taps as shown in Fig. 170.

The switch for operating the starting compensator and motor is the same as that used on the three-phase service.

Other designs are used, one of which operates the compensator as follows: For starting the motor the switch handle moves from the off position to the first starting position, where a low voltage is applied to the motor; then to a second starting position where a higher voltage is applied; and then to the running position, where the motor is connected directly across the line, the compensator being disconnected from the circuit. For stopping, the switch handle is moved to a notch still further along than the running position, the movement of the switch handle being in the same direction as in starting. In the latter position the switch handle is released to that it can be moved back to the off position ready to start again.

The other form of compensator used to indicate the variations of voltage at the point of distribution under all conditions of load without appreciable error between no-load and overload, consists of three parts: a series transformer, a variable reactance, and a variable resistance. The compensator is adjusted to allow for the resistance and inductive reactance of the line. If these are properly adjusted, a local circuit is obtained corresponding exactly with the line circuit, and any change in the line produces a corresponding change in the local circuit, causing the voltmeter always to indicate the potential at the end of the line or center of distribution, according to which is desired. It is well known that the drop in a direct-current circuit is dependent upon the resistance, but in an alternating-current circuit it is due not only to the resistance of the lines, but also to the reactance. The reactance usually causes the drop to be greater than it would if the resistance were the only factor. Therefore, it is necessary that a compensator should give accurately the voltage at the load at all times, whatever may be the current and power-factor.

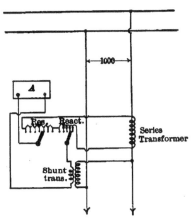

Fig. 171.—Form of compensator used to indicate the variation of voltage at the point of distribution under all conditions of load.

Fig. 171 shows a series transformer in series with the line; and having, therefore, in its secondary circuit a current always proportional to the current in the line. The reactors and resistors are both so wound that any proportion of the winding can be cut in or out of the voltmeter circuit, so modifying the reading of the station voltmeter that it corresponds with the actual voltage at the point of consumption, regardless of the current, power-factor, reactance, and resistance in the line. For balanced two- and three-phase circuits one compensator is sufficient.

In adjusting this type of compensator, it is advisable to calculate the ohmic drop for full-load and set the resistance arm at the point which will give the required compensation and then adjust the reactance arm until the voltmeter reading corresponds to the voltage at the point or receiving station selected for normal voltage. This compensator is commonly called the "Line-drop compensator."

Its connections proper, as used in practice at the present time are shown in Fig. 172. Each line of which the voltage at the center of distribution is to be indicated or recorded in the station, must be provided with a voltmeter as shown in Fig. 172, No. 1, which must be adjusted for the ohmic and reactive drop of each line respectively. For example, take such a line giving the factors shown in Fig. 173; where R is ohmic resistance of 10, X_s is the reactance of 10 ohms, and E is the load voltage. To fulfil the conditions of $E = 100$, it is necessary that the voltage of supply or generator voltage be increased to nearly 110.5, and the voltmeter No. 2 of Fig. 160 will indicate this value. With unity power factor it will be noted that the total live drop is due almost entirely to the line resistance R, and is practically independent of the line reactance.

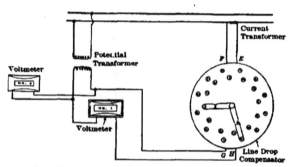

Fig. 172.—Connections for live-drop compensator.

As the power factor of the line decreases, the effective voltage produced by the reactance increases, until at an imaginary zero power-factor load the total drop is due almost entirely to the reactance.

The voltage diagram of such a circuit is shown in Fig. 174, where R and X_s and E are as before, but the line has an 80 per cent. power-factor load. The vector line E_g in this case represents 114 which gives $E = 100$. For the above or any other live conditions for which the compensator is set it is correct for all loads; for, as the drop in voltage in the line decreases, due to decreasing load, the voltage drop in the compensator decreases in equal measure. For other power factors a simple adjustment will be incorrect.

Automatic regulation of single-phase feeders presents no

difficulties, in that there is but one definite point to regulate and the "boost" or "lower" of the regulator is directly added to or subtracted from the voltage of the feeder. If regulation at the station is desired, only a potential transformer is necessary —if regulation for compensation of drop at some distant point is desired, a series transformer connected in series with the feeder is added.

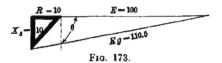

Fig. 173.

It very often happens that one phase of a three-phase feeder is used for lighting and a single-phase regulator installed. In making such an installation, the regulator must have its seconday winding in series with the line, its primary being connected across the phase. Now, if the load on this feeder is purely lighting, the power factor would remain constant and approximately 100 per cent., but should the power factor vary considerably,

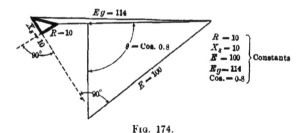

Fig. 174.

it may cause the current to be out of phase with the voltage to such an extent that satisfactory compensation may not be obtained. For such service the best arrangement would be to use two cross-connected series transformers, one connected in series with each conductor as, for instance, A and B of the phase across which the primary of the regulator is excited. With this connection (see Fig. 175) the line drop compensator is set to compensate for ohmic and inductive drop to the load center and the voltage will automatically be maintained at the desired value irrespective of changes in load or power factor.

224   STATIONARY TRANSFORMERS

Hand regulation has long been replaced with automatic regulators. With the development of the automatic regulator

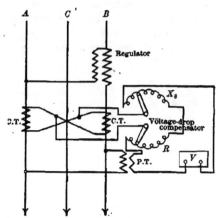

FIG. 175.—Single-phase regulator and voltage-drop compensator connections to a three-phase system.

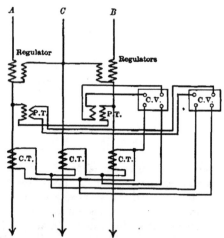

FIG. 176.—Two single-phase regulators on three-phase system, showing the use of the contact making voltmeter.

there has followed a perfection of the contact-making voltmeter, which displaces the operator and regulates the voltage automatically. This instrument is composed of a solenoid with two windings; a shunt winding which is connected in parallel with the secondary of a potential transformer, and a series winding (differential with respect to the shunt winding) which is connected in series with the secondary of the series transformer, the primary of which is in series with the feeder circuit. A movable core passes through the center of the solenoid, and to the top of this core is attached a pivoted lever. The lever carries at its other end a set of contacts which make contact with an upper and a lower stationary contact. The lever is set by means of a spring acting against the core, so that its contacts are midway between the upper and lower stationary contacts when normal voltage is on the shunt coil of the meter. The stationary contacts form, when closed, a circuit to one or the other of two coils of a relay switch, which in turn controls a motor on the regulator cover. Any deviation of voltage from normal causes contact to be made and the regulator corrects for this change, bringing the voltage back to normal. When compensating for line drop to a distant point the current coil is used and as the load increases, the regulator boosts the voltage by the proper amount. In this manner the meter can be set so that constant voltage can be maintained at a great distance from the regulator.

The connections of this instrument for a feeder circuit where lighting is connected on only two phases of a three-phase system and motors connected to the same, are shown in Fig. 176. Two single-phase regulators and three series transformers (one transformer in each conductor) are necessary. If the series transformer in the middle conductor or phase were not installed, proper compensation could not be secured, owing to both phase displacement and an unbalancing of current in the three phases.

Three single-phase regulators are usually employed when lighting and power are taken from all the three phases of a three-phase system. One three-phase regulator may be used, in which case it is better to employ only two series transformers cross-connected, so as to get an average of unbalanced current. With single-phase regulators as shown in Fig. 177 each phase can be adjusted independently of the others and constant voltage established at the load center of each. For this reason it is better to use three single-phase regulators in preference to one three-phase regulator.

For three-phase four-wire systems, three single-phase regulators are employed having their secondaries connected in series with a phase conductor and their primaries excited from phase conductors to neutral. This method is equivalent to three independent single-phase circuits.

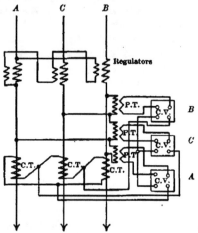

Fig. 177.—An installation giving perfect regulation when properly installed and operated.

CHAPTER XIV

TRANSFORMER TESTING IN PRACTICE

In order to determine the characteristics of a transformer the following tests are made:
1. Insulation.
2. Temperature.
3. Ratio of transformation.
4. Polarity.
5. Iron or core loss.
6. Resistances and $I^2 R$.
7. Copper loss and impedance.
8. Efficiency.
9. Regulation.
10. Short-circuit test.

Insulation.—The insulation of commercial transformers should be given the following tests:

a. Normal voltage with overload.

b. Double voltage for 30 minutes and three times the normal voltage for five minutes. (Distribution 2000 to 6000 volt transformers only.)

c. Between primary, core and frame.

d. Between primary and secondary.

e. Between the secondary, core and frame.

The National Board of Fire Underwriters specify that the insulation of nominal 2100-volt transformers, when heated, should withstand continuously for one minute a difference of potential of 10,000 volts alternating current between the primary and secondary coils and the core.

For testing the insulation of transformers, a high-potential testing set with spark-gap is required. The testing set should, preferably, have low reactance so that the variation in voltage, due to leading and lagging currents, will not be large. The voltages will be practically in the ratio of the turns, and the high-tension voltages may be determined by measuring the low-tension voltages and multiplying by the ratio of transformation.

STATIONARY TRANSFORMERS

Where a suitable electrostatic voltmeter is available the high-tension voltage is obtained by direct measurement.

In applying insulation tests, it is important that all primary terminals should be connected together as well as all secondary terminals, in order to secure a uniform potential strain throughout the winding. In testing between the primary and secondary or between the primary and core and frame, the secondary must be connected to the core and frame, and grounded.

In making the test, connect as shown in Fig. 178. The spark-gap should be set to discharge at the desired voltage, which may be determined directly by means of test with static voltmeter, or by the spark-gap table giving sparking distances in air between opposed sharp needle-points for various effective sinusoidal voltages in inches and in centimeters.

TABLE IX.—SPARKING DISTANCES

Table of Sparking Distances in Air between Opposed Sharp Needle-Points, for Various Effective Sinusoidal Voltages, in inches and in centimeters.

Kilovolts (Sq. root of mean square)	Distance		Kilovolts (Sq. root of mean square)	Distance	
	Inches	Cms.		Inches	Cms.
5	0.225	0.57	140	13.95	35.4
10	0.47	1.19	150	15.0	38.1
15	0.725	1.84	160	16.05	40.7
20	1.0	2.54	170	17.10	43.4
25	1.3	3.3	180	18.15	46.1
30	1.625	4.1	190	19.20	48.8
35	2.0	5.1	200	20.25	51.4
40	2.45	6.2	210	21.30	54.1
45	2.95	7.5	220	22.35	56.8
50	3.55	9.0	230	23.40	59.4
60	4.65	11.8	240	24.45	62.1
70	5.85	14.9	250	25.50	64.7
80	7.1	18.0	260	26.50	67.3
90	8.35	21.2	270	27.50	69.8
100	9.6	24.4	280	28.50	72.4
110	10.75	27.3	290	29.50	74.9
120	11.85	30.1	300	30.50	77.4
130	12.90	32.8			

After every discharge the needle points should be renewed. The insulation test which should be applied to the winding of a transformer depends upon the voltage for which the transformer is designed. For instance, a 2100-volt primary should withstand a difference of potential of 10,000 volts, and a 200-volt secondary should therefore be tested for at least 2000 volts. The length of time of the insulation test, varies with the magnitude of the voltage applied to the transformer, which, if severe, should not be continued long, as the strain may injure the insulation and permanently reduce its strength.

Transformers are sometimes tested by their own voltage. One side of the high-tension winding is connected to the low-tension winding, and the iron, and the transformer operated at a voltage above the normal to give the necessary test voltage. The same test is repeated when the other end of the high-tension winding is connected and the one side disconnected.

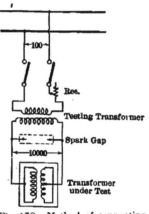

Fig. 178.—Method of connecting apparatus for insulation test.

In making insulation tests great care should be taken to protect not only the operator but others adjacent to the apparatus under test. If it is necessary to handle the live terminals, only one should be handled at a time, and whenever possible, it should be insulated beyond any possibility of the testing set being grounded.

Another insulation test called "over-potential test" is made for the purpose of testing the insulation between adjacent turns and also between adjacent layers of the windings. In applying the over-potential test, the exciting current of a transformer is always increased.

This test usually consists of applying a voltage three to four times the normal voltage to one of the windings with the other winding open-circuited. If this test is to be made on a 2000-volt winding, at three times its normal voltage, 6000 volts may be applied to one end of the winding in question, or:

3000 volts to a 1000-volt winding,
1200 volts to a 400-volt winding,
 300 volts to a 100-volt winding.

In general, this test should be applied at high frequency so that the exciting currents referred to above may be reduced. The higher the frequency the less will be the amount of current required to make the test. It is recommended that 60 cycles be the least used, and for 60-cycle transformers 133-cycle currents be applied, and for 25-cycle transformers 60 cycles.

The highest voltage transformer built up to the present time for power and industrial purposes was tested at 280,000 volts, or double voltage for which it was designed.

Double voltage is applied to test the insulation between turns and between sections of coils, these being the cause of practically 85 per cent. of transformer burnouts. A better and surer test than this would be to apply twice the normal voltage for one minute followed by another test for five minutes at one and one-half times normal voltage. The latter test is to discover any defect that may have developed during the double-voltage test, and yet not have become apparent in the short time the double voltage was applied.

The application of a high voltage to the insulation of a transformer is the only *real* method of determining whether the dielectric strength is there. Mechanical examination of the keenest kind is false, and measurement of insulation resistance is not very much better; since insulation may or may not show resistance when measured by a voltmeter with *low voltage*, but offer comparatively little resistance to a *high-voltage* current.

In working the high-voltage test between the primary and the core or the secondary, the secondary should always be grounded for the reason that a high voltage strain is induced between the core and the other winding which may be greater than the strain to which the insulation is subjected to under normal operation, and, of course, greater than it is designed to stand constantly. When testing between the primary or high-voltage side of, say, a high ratio transformer, and the core, the induced voltage strain between the low-voltage winding and core may be very high and the secondary may be broken down by an insulation test applied to the high-voltage side under conditions which would not exist during normal operation of the transformer. The shorter the time the voltage is kept

TRANSFORMER TESTING IN PRACTICE

on, with correspondingly higher voltage to get the desired severity of test, the less will be the deterioration of the insulation. Every transformer should be tested with at least twice its rated voltage, the reason for this being necessary is because of the many abnormal conditions of operation which occur. On a very high voltage system, if one side of the winding becomes grounded, the whole rated voltage is exerted between the winding and the iron core; and sometimes during normal operation (so far as exterior observtion indicates) a difference of potential will occur, lasting but a small fraction of a second or minute, which might be as high as the testing voltage required.

Practically all high-voltage transformers are wound with copper strip or ribbon, one turn per layer, the coils being insulated uniformly throughout excepting the end turns which are sometimes insulated to withstand voltages of 5000 volts between turns. Take the case of a transformer with a normal voltage between turns of 80 volts, it would mean that in applying the *standard voltage test, i.e.*, twice the rated high-tension voltage between the high-voltage winding and low-voltage winding connecting the latter to the core we shall receive 220,000 volts across a 110,000-volt winding, or an induced voltage of 160 volts between turns.

Several years ago it was agreed to lower the standard high-voltage test specifications to 1.5 times the rated voltage, and all sorts of breakdowns happened until a change was made to the higher test. The double voltage test is not too high and not too severe a test no matter how high the rated voltage of the transformer might be for power and industrial purposes. For commercial testing transformers it will not apply. Such a transformer was recently made for 400,000 volts and before it was shipped from the factory it was given a full *half-hour* test at 650,000 volts with the center of the high-voltage winding grounded. Since then one has been made for 750,000 volts. About the highest test made so far, on commercial power transformers was that of a 14,000 kv-a 100,000 volt transformer which was given 270,000 volts across its high-voltage winding to all other parts. This transformer was for a 60-cycle system and covered a floor space of 23 ft. ×8.5 ft., it being 18 ft. high. Another transformer of 10,000 kv-a at 70,000 volts operating on a 25-cycle system was give a high-voltage test of 180,000 volts. The former transformer was fitted with oil-filled terminal bushings and during the test of these bushings at 270,000 volts, no corona was visible even when in utter darkness. The latter transformer was fitted with the condenser type of terminal bushings which is made up of alternate layers of insulating and conducting material.

In the operation of transformers it has been found that we are confronted with the difficult problem of taking care of the voltage rise between turns that may not increase the line voltage sufficient to be noticeable but be so high in the transformer itself and only effective across a few turns as to short-circuit and burn-out that part where the excessive voltage is concentrated. In fact it is quite possible to get 100 times normal voltage across a small percentage of the total turns and at the same time have no appreciable increase in voltage at the terminals of the transformer.

With or without extra insulation on the end turns of transformers operating on high-voltage systems, the voltage difference on the end turns due to switching, etc., is there, and the only way out of the difficulty is to provide sufficient insulation to make it safe irrespective of the external choke coils which are always provided.

This double voltage test is made after the transformer has been thoroughly dried out and the quality of oil brought up to standard. In the past some makers placed too much reliance on oil as an insulator and consequently left out much solid insulation; the result was that many burnouts occurred, because the oil could not be relied upon always, its insulating proportions decreasing with age. In many cases burnouts occurred immediately the transformers were put into service after the drying-out process had been completed, and the voltage slowly brought up to the desired value.

Temperature.—The temperature or *heat test* of a transformer may be applied in several ways, all of which are arranged to determine as nearly as possible the working temperature conditions of the transformer in actual service.

Before starting a temperature test, transformers should be left in the room a sufficient length of time for them to be affected alike by the room temperature.

If a transformer has remained many hours in a room at constant temperature so that it has reached approximately uniform temperature throughout, the temperature of the surface may be taken to be that of the interior, or internal temperature. If, however, the transformer is radiating heat to the room, the temperature of the surface will be found to give little indication of the temperature of the interior.

To ascertain the temperature rise of a transformer, thermometers are sometimes used, which give only comparative results in temperature and such measurements are, therefore, useful only in ascertaining an increase in temperature during the heat run. If thermometers are used they should be screened from local air-currents and placed so that they can be read without being removed. If it is desired to obtain temperature curves, thermometer readings should be taken at half-hour intervals throughout the test and until the difference between the room temperature and that of the transformer under test is constant.

In order to determine temperature rise by measurement of resistance, it is necessary to determine first what is called "cold" resistance by thermometer measurements after the transformer has remained in a room of constant temperature for a sufficient length of time to reach a uniform temperature throughout its windings.

The temperature rise by resistance gives the average rise throughout the windings of the transformer, and to obtain average temperature rise of each of the windings, separate resistance readings should be taken of each.

The temperature rise by means of resistance may be determined by the use of the following equation:

$$R = R_0(1 + 0.004t); \qquad (31)$$

or by equation

$$\text{Resistance at } S°C = \frac{(238 \times S)}{(238t)} R' \qquad (32)$$

where R' is the resistance at any temperature t.

where R_0 is the resistance at room temperature; R the resistance when heated, and t the rise in temperature. The temperature coefficient of resistance is taken at 0.004 as 25° C. Considering the above equation, the temperature rise corrected to 25° C. may be determined in the following manner.

Example: Let room temperature be 20° C., and absolute temperature of transformer 60° C. Ascertain correct temperature rise.

The temperature is apparently $60 - 20 = 40°$ C., but since the room temperature is 5° lower than the standard requirements, a correction of $0.5 \times 5 = 2.5$ per cent. must be added giving a corrected temperature of

$$\frac{100 \times 2.5 \times 40}{100} = 41° \text{ C.}$$

Thus with a room temperature of 20° C. the rise in temperature calculated from the above equation should be added by 2.5 per cent., or with a room temperature of 35° C., the rise in temperature should be decreased by 5 per cent.; and with a room temperature of 15° C., the rise in temperature should be increased by 5 per cent., and so on.

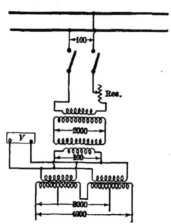

Fig. 179.—Method of connecting transformers and instrument for an over-potential test.

If the room temperature differs from 25° C. the observed rise in temperature should be corrected by 0.5 for each degree centigrade. This correction is intended to compensate for the change in the radiation constant as well as for the error involved in the assumption that the temperature coefficient is 0.004, or more correctly, 0.0039, remains constant with varying room temperatures.

To measure the increase of resistance let us take the following example. The primary resistance of a certain transformer is 8 ohms, and at its maximum operating temperature, 9 ohms. Temperature of room during test is 30° C. Ascertain corrected temperature rise.

The primary resistance taken at a temperature of 30° C., when referred to temperature coefficient of 0.4 per cent. per degree, represents a rise of $30 \times 0.4 = 12$ per cent. above its value at zero centigrade, which is

$$\frac{8 \times 100}{112} = 7.14 \text{ ohms.}$$

The maximum operating temperature of 10 ohms represents a rise of

$$\frac{(9 - 7.14)\ 100}{7.14} = 26.05 \text{ per cent.}$$

above value at zero, and is equal to

$$\frac{26.05}{0.4} = 65° \text{ C.}$$

absolute temperature.

Deducting from this the room temperature at 30°, the apparent rise $= 35°$. C. Since the room temperature during test was 5° above standard requirements, a correction of 0.5×5, or 2.5 per cent. must be substracted, giving a corrected rise of

$$\frac{35 \times 90}{100} = 3.15° \text{ C.}$$

It is well known that high temperatures cause deterioration in the insulation as well as increase in the core loss.

The average temperature rise of the coils of transformers can be more accurately determined from the hot and cold resistances, and is calculated as follows:

$$\text{Temp. rise (°C)} = (238.1 + T)\left(\frac{\text{hot Res.}}{\text{cold Res.}} - 1\right) \qquad (33)$$

$T =$ temperature at which cold resistance is taken.

As already stated above, if the final room temperature is less than 25° C., the temperature rise should be corrected by adding 0.5 of 1 per cent. for each degree less; if room temperature is greater than 25° C., subtract 0.5 of 1 per cent., for each degree greater. While these readings are being taken the exciting current and the load current are held constant and the thermometers are read at 30 or 60 minute intervals for a period of several hours and until the constant temperature has been reached. The thermometers should be placed in the room, in the oil, on the cores, tank and various other parts of the transformer when possible.

A method of heat-run used to some extent, and known as the *"Opposition"* test, is shown in Fig. 180. In this test two transformers of the same capacity, voltage and frequency are required and connected as shown in diagram. The two secondary windings are connected in parallel, and the two primary windings connected in series in such a way as to oppose each other. The two secondary leads receive exciting current at the proper voltage and frequency, while the primary leads receive a current equal to the desired load current; the wattmeter in the primary circuit

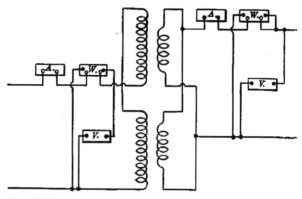

FIG. 180.—Method of connecting apparatus for heat test, known as "opposition" test.

measures the total copper loss, and that in the secondary the total core loss.

Another method often used and called the "motor generator test" is shown in Fig. 181. In this test two transformers are used, having their high-tension windings connected together. Proper voltage is applied to the low-tension winding of one of the transformers, and the low-tension winding of the other transformer is connected to the same source. Then with the switch s open, the wattmeter reads the core losses of both transformers, and with s closed, it reads the total loss. Subtracting the core loss from the total, the copper loss is obtained. This method requires (as is also the case in the opposition test) that only the losses be supplied from the outside.

TRANSFORMER TESTING IN PRACTICE 237

At the present time the flow point of the impregnating compounds gives a temperature limit of about 90° C. It is possible that the development of synthetic gums will soon reach a stage to permit of actual operating temperatures of at least 125° to 150° C. The only difficulty with such an operating temperature will be with the oil.

Certain practices of drying out transformers are applicable to temperature tests (see Chapter IX). In drying out trans formers it is always more convenient to short-circuit the low-voltage winding and impress sufficient voltage on the high-

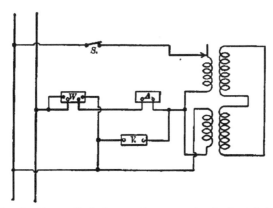

Fig. 181.—Another method of connecting apparatus for heat test known as "motor generator" test.

. voltage winding to cause about 20 to 30 per cent. current to flow through the transformer coils. This current is found quite sufficient to raise the *temperature* of the coils to the desired limit.

To make normal current flow through the windings when the secondary is short-circuited requires a voltage of about 3.3 per cent. of the high voltage winding, according to the way the windings are connected (series or parallel). For example: It is desired to dry out a 100,000 volt, 10,000 kw. single-phase transformer; a 5000 kw., a 2500 kw., and a 1250 kw. of the same voltage, etc. What will be the voltage necessary to circulate 20 and 30 per cent. of normal current through the coils of the transformers? The answer to this is best given in the following table:

TABLE XI

Conditions of test	Capacity of transformers in kw.			
	10,000	5000	2500	1250
Normal high-voltage current at 100,000 volts on full-load.	100	50	25	12.5 amp.
Voltage required to circulate same when low-voltage winding is short-circuited.	3300	3300	3300	3300 volts
At 20 per cent. normal current....	20	10	5	2.5 amp.
Voltage required for 20 per cent...	660	660	660	660 volts
At 30 per cent. normal current.....	30	15	7.5	3.75 amp.
Voltage required for 30 per cent...	990	990	990	990 volts

If one is obliged by circumstances to short-circuit the high-voltage winding, the same per cent. voltage holds good. Assuming the low voltage to be 5000 volts, it will require 33 volts for 20 per cent. normal current, and 50 volts for 30 per cent. normal current. With the high-voltage winding connected for 50,000 volts, the current values given above will be doubled but the temperature conditions will remain about the same.

The maximum values of temperature for large transformers is a rise of 40° C. under rated load, this value having for its basis a room temperature of 25° C. For rated overloads the limiting temperature rise is 55° C. In an earlier part it has been stated that, in general, it is found that transformers will operate quite satisfactorily when worked at their limiting temperatures; that is to say, around that point where the best efficiency and full-load is obtained. A condition of this kind might mean that for short periods of time an overload is required, hence a high ratio of copper and iron loss, and decreased first cost of transformers.

High temperatures are objectionable in transformers. Their effect on the insulation at temperatures about 100° C. means gradual deterioration; their effect on the copper loss is a decided objection, this loss increasing about 10 per cent. with an increase of 25° C. in the temperature; their effect on the oil is to increase the deposition of hydrocarbon on the windings and internal cooling apparatus, and a further bad effect is their tendency to increase the "aging" of the iron (not including the present improved silicon steel).

To get at, and measure, the maximum temperature affecting the insulation is almost impossible and only the average temperature is measured during a regular test, the temperature being taken at the top of the oil and not at the immediate surface of contact with the coils or core of the transformer.

The Standardization Rules of the A.I.E.E. state that the rise in temperature of a transformer should be based on the temperature of the surrounding air. The cooling medium for oil-insulated self-cooled and for forced-air cooled transformers is the surrounding air, but for oil-insulated water-cooled transformers the cooling medium is water, and the temperature rise to be considered in this type is that of the ingoing water and not the temperature of the surrounding air. On this basis a transformer of this type will be about 10° C. less than one specified on the basis of the temperature of the surrounding air.

The class of oil used for insulating purposes has a great effect on the temperature. An increase in the viscosity of the oil means an increase in the frictional resistance to its flowing. The velocity of circulation is reduced, thereby causing an increased rise in the temperature.

As yet it has not been possible to formulate a correct theory of the laws of cooling for general cases which will indicate once for all that combination of conditions which is most favorable to cooling, and enable one to say with considerable accuracy not only what will be the average temperature rise in any given case, but also what will be the maximum rise.

Ratio of Transformation.—The ratio of a transformer is tested when the regulation test is made. It is the numerical relation between the primary and secondary voltage. The ratio of a transformer must be correct, otherwise the service will be unsatisfactory, because the secondary voltage will be too high or too low.

For successful parallel operation, correct ratios are essential; otherwise cross-currents will be established through the windings.

A method of ratio test is shown in Fig. 182, where the primary of the transformer under test is in parallel with the primary of the standard ratio transformer, and the two secondary windings are connected in series.

Standard transformer ratios are usually an exact multiple of 5 or 10.

TABLE XII.—TABLE OF TEMPERATURE COEFFICIENTS OF RESISTIVITY IN COPPER AT DIFFERENT INITIAL TEMPERATURES CENTIGRADE

Initial temperature cent.	Temp. coefficient in per cent. per degree cent.	Initial temperature cent.	Temp. coefficient in per cent. per degree cent.
0	0.4200	26	0.3786
1	0.4182	27	0.3772
2	0.4165	28	0.3758
3	0.4148	29	0.3744
4	0.4131	30	0.3730
5	0.4114	31	0.3716
6	0.4097	32	0.3702
7	0.4080	33	0.3689
8	0.4063	34	0.3675
9	0.4047	35	0.3662
10	0.4031	36	0.3648
11	0.4015	37	0.3635
12	0.3999	38	0.3622
13	0.3983	39	0.3609
14	0.3967	40	0.3596
15	0.3951	41	0.5383
16	0.3936	42	0.3570
17	0.3920	43	0.3557
18	0.3905	44	0.3545
19	0.3890	45	0.3532
20	0.3875	46	0.3520
21	0.3860	47	0.3508
22	0.3845	48	0.3495
23	0.3830	49	0.3482
24	0.3815	50	0.3471
25	0.3801		

TRANSFORMER TESTING IN PRACTICE 241

Low-voltage distribution transformer ratios are (low-voltage windings) 110, 220, 440 or 550 volt to (primary voltage winding) 1,110/2,200, 3,300, 6,600 and 10,000 volts.

High-voltage transformers are wound for 11,000, 22,000, 33,000, 44,000, 66,000, 88,000, 110,000 and 140,000 volts.

Occasionally transformers are required with ratio-taps on the primary winding so that they may be operated at the maximum, intermediate or minimum ratio.

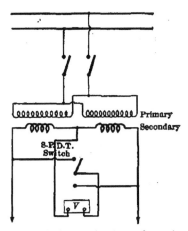

FIG. 182.—Method for ratio of transformation test.

The advantages of such taps are (a) voltage compensation due to line drop; (b) the possibility of operating the complete system on any of the intermediate ratios, (assuming neutralized system.)

Both of these advantages are sometimes desirable but it would be better to obtain them by other means than cutting down the normal rating of the transformer.

It is evident that if the primary voltage is maintained constant while operating on any of the intermediate taps, the transformer is operated at a greater voltage per turn and therefore at a greater iron loss than when the total winding is used. The copper loss is reduced somewhat so that the total full-load losses are not materially increased. The all-day efficiency is reduced very materially since the iron loss exists for 24 hours of the day and the copper loss only about three to four or five hours. And, as regards compensating for excessive line drop, it is true that if the transformer is connected for a lower ratio and at such places as have excessive line drop, the decreased primary voltage impressed upon the decreased primary turns will produce approximately normal core loss and the desired secondary voltage during the period of full-load on the system. However, during light load, when a heavy current is no longer in the primary and an excessive drop no longer exists, the transformer connected and operating on any of the intermediate or lower ratio taps will be subjected to full primary voltage impressed on the reduced primary turns, and the core loss of the transformer will become excessive and the secondary voltage increased to a dangerous limit, that is to say, dangerous so far as burn-outs of incandescent lamps are concerned, for, whoever should happen to turn on their lamps during the period of light-load operation, is sure to suffer from excessive lamp burn-outs. Operation at 10 per cent. above normal voltage for which the lamp is designed, reduces its life to 15 per cent. of its normal value. Consequently by operating a transformer on its intermediate ratio-taps, offers the disadvantage of excessive core loss for at least 20 to 21 hours in the day, and excessive burning out of lamps during the same period of time.

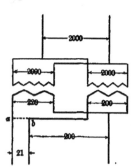

Fig. 183.—Effect of connecting two transformers of different ratios in parallel.

Decreasing the ratio of a transformer from, say, 10–1 to 9–1 and maintaining the voltage constant, increase the core loss approximately 20 per cent.

A difference of about 10 per cent. in ratios of primary and secondary voltages will result in a circulation of about 100 per cent. full-load current. And a difference of 2 per cent. in the ratios may result in a circulation of 20 per cent. full-load current, or a 1 per cent. difference in ratio may result in a 10 per cent. circulation of full-load current; thus showing the absolute necessity for having the ratios always exactly the same. For example; take Fig. 183 and assume that the percentage impedance volts of each transformer is 5 per cent., and the measured difference between points a and b shows 21 volts. What will be the circulating current with open secondaries?

The 21 volts is effective in circulating current through the transformer windings against the impedance of the transformers, the amount of current being expressed in per cent., as

$$I \text{ per cent.} = \frac{k \times 100}{Z^\circ} \qquad (34)$$

where k is the difference in voltage ratio and Z° the total impedance volts of the two transformers.

The circulating current in this case will be 100 per cent., or

$$I^\circ = \frac{k \times 100}{Z^\circ} = \frac{21 \times 100}{21} = 100$$

Taking the transformers at 2000 kv-a each, it is evident that there will be 10,000 amp. flowing through the secondary windings of one transformer and 9100 amp. through the secondary windings of the other when a and b are connected together.

Before connecting any two transformers in parallel it is advisable first to measure the voltage difference between a and b points shown in Fig. 183.

Polarity.—The most simple method for testing the polarity is to connect the primary and secondary windings of the transformers in parallel, placing a fuse wire in series with the secondary winding. If the transformers are of opposite polarity the connection will short-circuit the one transformer on the other, and the fuse will blow. Many burnouts are due to wrong connections of this kind.

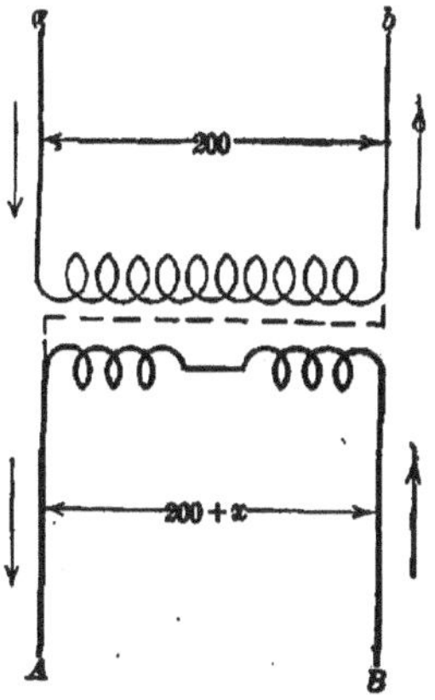

Transformers are generally assembled so that certain selected leads are brought out the same in all transformers of the same type. See Fig. 184.

Fig. 184.—Simple method for testing the polarity of transformers.

The primary terminal (a) should be of opposite polarity to the secondary terminal (A). If we apply 200 volts to the primary, $a\,b$, of the transformer, the voltage between $a\,B$ should be greater than the voltage applied to $a\,b$, if the transformer is of the correct polarity, or less if of opposite polarity; that is to say, if two single-phase transformers, both of *positive* polarity or both of *negative* polarity, are to be operated in parallel, they should be connected together as shown. If all transformers are alike, they may have the same polarity, but if some are of different designs or are made by different manufacturers, their polarity may be different.

Single-phase polarity is very easily determined; not so with polyphase transformers since both phase relation and rotation

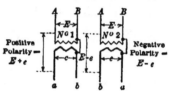

FIG. 185.—Positive and negative polarity of single-phase transformers.

must be considered; in fact polyphase polarity may mean a large number of possible combinations.

Considering first, the test for polarity of single-phase transformers, it is best to consider the direction of voltages to know whether they are in phase or in opposition, that is, 180 degrees out of phase. *Positive polarity* means that if, during the test, $A-b$ is the sum of voltage $A-B$ and $a-b$, positive polarity is ob-

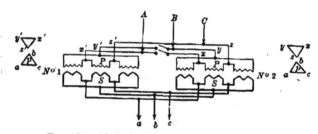

FIG. 186.—Method of finding threee-phase polarity.

tained. *Negative polarity* means that if, the voltage between $A-b$ is the value of secondary voltage less than $A-B$ and $a-b$, negative polarity is obtained. See Fig. 185.

In Fig. 185 we have:

Positive polarity = No. 1 transformer = $(A - B) + (a - b) = E + e$.
and
Negative polarity = No. 2 transformer = $(A - B) - (a - b) = E - e$.

This means that, in order to connect No. 1 and No. 2 in parallel, different leads must be connected together. *AA, BB* on one side and *aa, bb* on the other side. It is always better, however, when a positive polarity transformer is to be connected in parallel with a negative polarity transformer, to reverse the connections of either the high-voltage winding or the low-voltage winding of one of the transformers.

Fig. 186 shows the method of finding whether two three-phase transformers have the same polarity.

In making the above test similarly located terminals should be connected together as shown. If no voltage is indicated between leads x' and x or between y' and y, the polarities are the same and the connections can be made and if desired, put into regular operation. If, however, there is a difference of voltage between

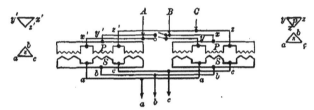

Fig. 187.—Testing for polarity in three-phase systems.

$y' - y$ or $x' - x$, or both, the polarity of the two groups is not the same and parallel operation is impossible. It is necessary to determine the polarity of each three-phase transformer separately.

The correct connections under these circumstances, are given in Fig. 187.

Iron or Core Loss.—The core loss includes the hysteretic and eddy-current losses. The eddy-current loss is due to currents produced in the laminations, and the hysteretic loss is due to molecular friction. The core loss remains practically constant at all loads, and will be the same whether measured from the primary or secondary side, the exciting current in either case being the same per cent. of the full load. The economical operation of a lighting plant depends in a large measure on the selection of an economical transformer. An economical transformer is seldom the one of lowest first cost, nor is it necessarily the one having the smallest full-load losses. It is the one which has the most suitable division of losses for the service for which it is to be used.

The hysteresis loss is dependent on the iron used, and in a given transformer varies in magnitude with the 1.6 power of the (induction) or magnetic density. An increase in voltage applied to a transformer causes an increase in core loss (see following table), while an increase in the frequency results in a corresponding decrease in core loss—the density varying directly as the voltage and inversely as the frequency. The eddy current loss varies in magnitude with the conductivity of the iron and the thickness of laminations. Both the hysteresis and eddy current losses decrease slightly as the temperature of the iron increases, and if the temperature be increased sufficiently the hysteresis loss might disappear entirely while the eddy current loss will show a decrease with increased resistance of the iron due to this temperature. Thus at full-load, or in other words, an increase in temperature to the limiting temperature rise of 40° C. may cause a decrease in core loss of about 5 per cent. depending on the wave form of the impressed voltage. For ordinary steel used in transformers, a given core loss at 60 cycles may consist of 72 per cent. hysteresis and 28 per cent. eddy current loss, the hysteresis loss decreasing with increased frequency while the eddy current loss is increased with increased frequency.

Low power factor of exciting current is not in itself very objectionable. This can best be explained by taking two transformers, one made up of ordinary iron and the other of modern silicon steel or "alloy-steel." Take, for example, two 5kw. transformers. The one made up of ordinary iron will have a core loss of about 64 watts while that made of silicon steel will have a core loss of only 45 watts; and, taking the exciting current of both to be 2 per cent. of the full-load current, we have the following power factors:

Transformer with ordinary iron:

$$\frac{64}{5000 \times 0.02} = 64 \text{ per cent. power factor}$$

and

Transformer with silicon steel:

$$\frac{45}{5000 \times 0.02} = 45 \text{ per cent. power factor}$$

which means that on no-load the one using ordinary iron has 19 per cent. better power factor.

As stated above, the lower the frequency the greater will be the iron loss. In ordinary commercial transformers a given core loss at 60 cycles may consist of 72 per cent. hysteresis and 28 per cent. eddy-current loss, while at 125 cycles the same transformer may have 50 per cent. hysteresis and 50 per cent. eddy-current loss. The core loss is also dependent upon the waveform of the applied e.m.f. A flat top wave gives a greater loss than a peaked wave and *vice versa*.

With a sinusoidal wave of e.m.f. applied on a transformer, the exciting current is distorted, due to the effect of hysteresis. If resistance is introduced into the primary circuit, however, the

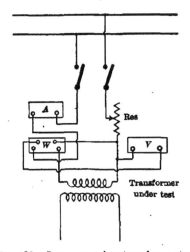

FIG. 188.—Iron or core loss transformer test.

exciting current wave becomes more sinusoidal and the generated e.m.f.-wave more peaked, the effect of these distortions tending to reduce the exciting current and core loss. Since the magnetic density varies with the voltage and inversely with the frequency, an increase in voltage applied to the transformer causes an increase in core loss, while an increase in frequency results in a corresponding decrease in core loss.

Of the several methods in use for determining core loss, the following method is the simplest to apply and gives very accurate results. See Fig. 188.

There are occasions when the core losses of a transformer are known while operating at a given frequency and voltage but when it is desired to correct these results for operation under other conditions other figures are necessary. In order to determine approximately the losses of a 60-cycle transformer when operating at other than rated voltage, the losses at rated voltage may be multiplied by the factors given in the following table:

TABLE XIII.—VARIATION OF CORE LOSS IN A 60-CYCLE TRANSFORMER WITH VARYING VOLTAGE

Operating voltage	Rated voltage of transformer						
	2000	2100	2200	2300	2400	2500	2600
	Losses						
2000	1.00	0.91	0.83	0.76	0.70		
2080	1.09	0.98	0.89	0.82	0.75	0.70	
2100	1.11	1.00	0.91	0.83	0.77	0.71	0.66
2200	1.23	1.11	1.00	0.91	0.84	0.77	0.72
2300	1.37	1.22	1.10	1.00	0.92	0.84	0.78
2400		1.35	1.21	1.10	1.00	0.92	0.85
2500		1.49	1.33	1.20	1.09	1.00	0.92
2600			1.46	1.31	1.19	1.09	1.00

The values given in the above table are only approximate because the variations with varying voltage depend largely upon the quality of steel and the density at which the transformers are operated.

Resistances.—The resistance of the primary and the secondary of a transformer may be determined by several different methods, the most common of which are "fall of potential" and "wheatstone bridge" methods. For commercial use the most satisfactory method is the fall of potential. In this method the resistance may be determined by Ohms law:

$$\text{Resistance} = \frac{\text{Volts}}{\text{Amperes}}. \quad (38)$$

TRANSFORMER TESTING IN PRACTICE

The measurement requires continuous current and a continuous-current voltmeter and ammeter. With the connection shown in Fig. 189, assume, for example, the ammeter reading to be 2.5 amperes, and voltmeter reading to be 11 volts. What is the resistance of coil?

The resistance of voltmeter used in test is 500 ohms, and the temperature of transformer coil is 30 degrees centigrade. Therefore, current taken by voltmeter at 11 volts is,

$$\frac{11}{500} = 0.022 \text{ amp.}$$

Current in transformer coil $= 2.5 - 0.022 = 2.478$ amp.

The ammeter reading includes the current in the voltmeter,

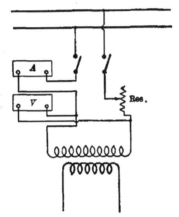

Fig. 189.—Method of finding the resistance of a transformer.

and in order to prevent error the resistance of the voltmeter must be much greater than that of the resistance to be measured.

Resistance of transformer coil at 30 degrees centigrade is,

$$\frac{11}{2.478} = 4.48 \text{ ohms.}$$

It is important that measurements be taken as quickly as possible, especially if the current be near the full-load values, and it is equally important in all cases that the voltmeter needle be at rest before the observation is taken, otherwise the values obtained will not be reliable. It is possible to have a current of sufficient strength to heat the coil so rapidly as to cause it to reach a constant hot resistance before the measurement is taken. The resistance of the transformer coil at 25° C., which is the temperature coefficient of 0.42 per cent. per degree from and at 0° C., is,

$$\frac{4.48 \times 100}{(0.42 \times 5) + 100} = 4.39 \text{ ohms.}$$

If the temperature of windings is different for each observation, then resistance must be calculated for each and the average taken. If the temperature of the windings is the same for all observations, then the average voltage and current may first be determined and the resistance calculated from the average values.

Copper Loss and Impedance.—When a transformer is delivering power, copper loss takes place, varying as the square of the

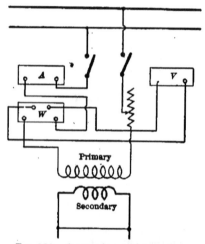

FIG. 190.—Copper loss and impedance.

current. It is due to the resistance of the windings and to the eddy currents within the conductors themselves.

The copper loss may be measured at the same time as the impedance-drop measurement by introducing a wattmeter as shown in Fig. 190. It may also be calculated from currents through conductors and resistance of conductors, as follows:

$$P = I_1^2 R_1 + I_2^2 R_2, \text{ in watts,}$$

wherein P is the power lost; I_1, the primary current; I_2, the secondary current; R_1 the primary resistance; and R_2 the secondary resistance.

The variation of copper loss with varying voltage on 2200 to 2600 volt 60-cycle transformers is given in the following table:

TABLE XIV.—VARIATION OF COPPER LOSS WITH VARYING VOLTAGE ON 60-CYCLE 2200 TO 2600 VOLT TRANSFORMERS

Varying voltage	Ratio voltage							
	2000	2080	2100	2200	2300	2400	2500	2600
	Losses							
2000	1.00	1.08	1.10	1.21	1.32	1.44	......	......
2080	0.98	1.00	1.02	1.12	1.22	1.33	1.44	......
2100	0.91	0.98	1.00	1.10	1.20	1.30	1.42	1.53
2200	0.83	0.89	0.91	1.00	1.10	1.19	1.29	1.40
2300	0.76	0.82	0.83	0.91	1.00	1.09	1.19	1.28
2400	......	0.75	0.77	0.84	0.92	1.00	1.09	1.17
2500	......	......	0.71	0.77	0.85	0.92	1.00	1.08
2600	......	......	......	0.72	0.78	0.85	0.93	1.00

In testing a three-phase transformer for copper loss and impedance the measurements can be made conveniently by connecting both the high-voltage and the low-voltage windings in delta and opening up any one corner of the delta on either the high-voltage or low-voltage side as desired and convenient for supply voltage, and inserting a wattmeter, voltmeter and ammeter and impressing sufficient *single-phase* voltage across this corner at the proper frequency to cause normal full-load current to flow through the windings. The wattmeter reading will give the copper loss, and this reading divided by the normal full-load input in the transformer will represent the per cent. copper loss. One-third of the voltage measured on the voltmeter divided by the normal voltage of the winding of one phase represents the percentage impedence drop.

The impedance in alternating-current circuits is similar to resistance in continuous-current circuits, that is to say, the expression

$$I = \frac{E}{R} = \text{Current} = \frac{\text{e.m.f.}}{\text{Resistance}} \qquad (42)$$

for continuous-current circuits is replaced in alternating-current circuits by the equivalent expression,

$$I = \frac{E}{\sqrt{R^2 + (X_s^2)}} = I = \frac{\text{e.m.f.}}{\text{Impedance}} = \frac{E}{R - jx} = \frac{E}{R + \sqrt{-1}\, Lw} \qquad (43)$$

where I is the current; E the impressed e.m.f.; X_s the inductive reactance; and R the resistance of the circuit.

The impedance of a transformer is made up of two components at right angles to each other. (Reactance and resistance.) It is expressed as

$$Z = \sqrt{R^2 + X^2} = (R - jx)R + \sqrt{-1}\, Lw \qquad (45)$$

Reactance may be inductive, X_s, or condensive X_c; this latter factor is never considered when dealing with transformers.

$$X_s = 2\pi f L \text{ and } X_c = \frac{1}{2\pi f C} = \frac{1}{\sqrt{-1}\, Cw} \qquad (46)$$

wherein f is the frequency in cycles per second; L is the inductance in henrys; and C is the capacity in farads.

The impedance of a transformer is measured by short-circuiting one of the windings, impressing an e.m.f. on the other winding and taking simultaneous measurements of voltage and current.

The impedance voltage varies very nearly with the frequency. In standard transformers the impedance voltage varies from 1 to 4 per cent., depending upon the size and design of the transformer.

Efficiency.—The efficiency of a transformer is the ratio of its net output to its input. The output is the total useful power delivered and the input is approximately the total power delivered to the primary; and consists of the output power plus the iron loss at the rated voltage and frequency, plus the copper loss due to the load delivered.

Example: Find the full-load, and half-load efficiency of a 5-kw., 2000 to 200-volt, 60-cycle transformer having an iron loss of 70 watts, a primary resistance of 10.1 ohms, a secondary resistance of 0.066 ohms.

The efficiency of the transformer under consideration is taken as follows:

Full Load:
- Primary $I^2 R$ .... 63 watts
- Secondary $I^2 R$ .... 42 watts
- Core loss .... 70 watts

- Total Losses .... 175 watts
- Output = ... 5,000 watts
- Input = 5,000 + 175 .... 5,175 watts
- Full load eff. $= \dfrac{5,000}{5,175} = 96.6$ per cent.

Half Load:
- Primary and Secondary .... 26 watts
- Core loss .... 70 watts

- Total Losses .... 96 watts
- Output = ... 2,500 watts
- Input = 2,500 + 96 .... 2,596 watts
- One-half load eff. $= \dfrac{2,500}{2,596} = 96.2$ per cent.

It will be noted that the iron loss remains constant at all loads but the copper loss varies as the square of the load current. The copper loss remains the same in all transformers of a given design and size, it is, therefore, only necessary to make these tests on one transformer of each rating and type.

The copper loss should preferably be determined from the resistances of the windings, rather than from the copper loss test by wattmeter. For other than full-load, the copper loss varies as the square of the load, the core loss remaining constant at all loads. The all-day efficiency takes into account the time during which these losses are supplied and is expressed as:

Per cent. all-day eff. =
$$\frac{100 + \text{watthours output}}{\text{w.-hrs. output} + \text{w.-hrs. copper loss} + \text{w.-hrs. core loss}} \quad (47)$$

The exact copper loss of a transformer must be known in order to calculate the efficiency. The core loss should be taken at exactly the rated voltage of the transformer and, when possible, with a sine wave current, otherwise considerable discrepancies may occur.

Regulation.—The regulation of a transformer with a load of given power-factor is the percentage of difference of the full load and no load secondary voltages with a constant applied primary voltage. It may be ascertained by applying full load to the transformer and noting the secondary voltage, then removing the load and noting the secondary open-circuit voltage.

The secondary voltage drop will be very much greater with an inductive load, such as induction motors or arc lamps, than it will be with incandescent lamps.

The regulation can be determined by direct measurement or calculation from the measurements of resistance and reactance in the transformer. Since the regulation of any transformer is only a few per cent. of the impressed voltage, and as errors of observation are liable to be fully 1 per cent., the direct method of measuring regulation is not at all reliable. By connecting the transformer to a circuit at the required voltage and frequency, using a lamp load or water rheostat on the secondary the regulation may be determined. This method, is, however, unsatisfactory, and much more reliance can be placed on the results of calculation.

Several methods have been proposed for the calculation of regulation, but the following is found quite accurate for inductive and non-inductive loads.

For inductive loads:

$$\% \text{ regulation} = \% X \sin \theta + \% I R \cos \theta. \qquad (48)$$
$$= \% E_x \sin \theta + \% E_r \cos \theta.$$

Per cent. E_x = the per cent. reactance drop.
Per cent. E_r = per cent. total resistance drop.
θ = the angle of lag of load current delivered.

Example.—Find the regulation of a transformer which has a reactance drop of 3.47 per cent. and a resistance drop of 2.0 per cent, when delivering a load to a circuit having a power-factor of 87 per cent.

The $\cos \theta = 0.87$ is 30 degrees. The sine of angle 30 degrees is 0.5. Then from the above formula:

Per cent. regulation $= 3.47 \times 0.5 + 2 \times 0.87 = 3.48$.

For non-inductive loads:

$$\% \text{ Regulation} = \% I R - \frac{\% IX_s - 2 IX_s i}{200} \qquad (49)$$

Per cent. $I R$ = per cent. resistance drop.

$\% X = \sqrt{\% \text{ impedance drop}^2 - \% \text{ resistance drop}^2} = \%$ reactance drop.

$i = \sqrt{\% \text{ exciting current}^2 - \% \text{ iron loss current}^2}.$

TRANSFORMER TESTING IN PRACTICE

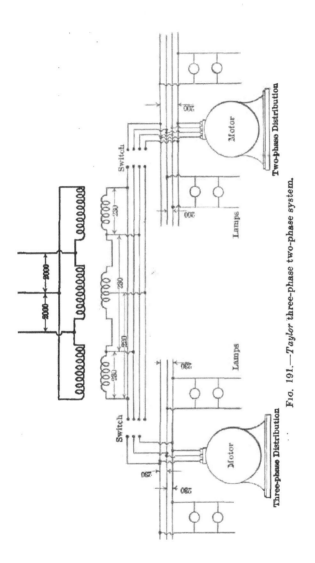

FIG. 191.—*Taylor three-phase two-phase system.*

For non-inductive load $\theta = 0$, $\sin \theta = 0$, $\cos \theta = 1$, we have, therefore:

$$\% \text{ regulation} = \% E_r$$

The above formula is practically correct for small values of angle θ, but the error becomes greater as θ increases.

Another simple and accurate method in use for calculating regulation is,

For non-induction loads:

$$\% \text{ regulation} = IR \frac{IX_s^2}{200} \qquad (50)$$

and for inductive loads:

$$\% \text{ regulation} = d + \frac{k^2}{200} \qquad (51)$$

where d is component drop in phase with the terminal voltage and k is the component drop in quadrature with the terminal voltage, IR is total resistance drop in per cent. of rated voltage and IX is reactance drop in per cent.

Take for example a 7.5 kw. 60-cycle single-phase transformer with a 10 to 1 ratio, secondary voltage at full load = 208 volts and

Sec. resistance = 0.0635 ohms. Primary resistance = 4.3 ohms.

Sec. $IR = 2.165$ volts = 0.985 %. Primary $IR = 14.65$ volts = 0.667 %.

and

$$IX = \sqrt{(\% \text{ impedance drop})^2 - IR^2} = 1.8 \%$$

For non-inductive load, using the formula above, we have

$$\% \text{ regulation} = 1.65 + \frac{1.8^2}{200} = 1.67 \%$$

and

For inductive load with a power factor of, say, 80%, we have

$$\% \text{ regulation} = 2.3 + \frac{0.46^2}{200} = 2.33 \%$$

where 2.3 is taken from formula, and is:—

$$d = w\, IX + IR \cos \theta = (0.6 \times 1.8) + (1.65 \times 0.8) = 2.3$$

and

0.46 is from formula, and is:—

$$k = w\, IR - IX \cos \theta = (0.6 \times 1.65) - (1.8 \times 0.8) = 0.46$$

where w is the wattless factor of load.

TRANSFORMER TESTING IN PRACTICE

Short-circuit Test.—For some time past it has been the practice of certain transformer-testing departments in America to subject certain types of transformers over a given k-v-a. capacity to short-circuit tests of from 5 to 25 times rated current.

It is now well known that the cause of many burn-outs is due to the large amount of power back of the transformers. It has also been shown that a certain milling occurs in transformers, and after repeated short-circuits the transformer breaks down, its coils being twisted in the shell type and displaced in the core type.

The short-circuit test is usually done at special times of the night so as not to affect the voltage regulation of the system. One winding of the transformer is connected to the power system (always many times the capacity of the transformers) and the other winding suddenly short-circuited. The tendency of the coils to flare out due to the excessive magnetic repulsion is the most important point of the test. This test is of very short duration, as the current sometimes reaches as high a value as 25 to 30 times full-load current.

CHAPTER XV

TRANSFORMER SPECIFICATIONS.

High-voltage transformer specifications are always interesting in that other more severe mechanical and electrical stresses have to be considered in their design, and the arrangement of coils, their form and make-up are so different. Below is given a winding specification of a high-voltage transformer which was specially made for an existing transformer operating at a lower voltage (the same iron being used over again).

WINDING SPECIFICATION FOR A WATER-COOLED OIL-FILLED 60-CYCLE 900-KW. 22,700-39,300 TO 2200-VOLT SINGLE-PHASE TRANSFORMER.

PRIMARY WINDING

Conductor cross-section, two 0.170 in. × 0.080 in. double cotton covered.

Weight, 750 lb. double cotton covered.

Inside Section.   *Outside Section.*

Turns $\begin{cases} 8 \text{ B.T. coils of 32 and 32} \\ 2 \text{ B.T. coils of 27 and 26} \end{cases}$ ... $\begin{cases} p_2, p_3, p_4, p_5, p_6, p_7, p_8, \\ p_9, p_1, \text{ and } p_{10}. \end{cases}$
618 in.

Winding taps made at end of fifth turn inside end of outside section p_5, p_6.

Insulation between turns (8300 ft.) 0.015 in. thick by 3/16 in. wide, consisting of two 0.005 in. hercules parchment and one 0.005 in. mica.

Reinforced Turn Insulation

26-turn section, last 12 turns (triple) turn insulated, and all turns 0.012 V.C.[1]

26-turn section, all other turns (double) turn insulated, and all turns 0.012 V.C.

26-turn section, all turns (double) turn insulated, and all turns 0.012 V.C.

[1] V.C. = varnish cambric.

TRANSFORMER SPECIFICATIONS

Coils p_1 and p_{10} special collars of 1 3/32 in. pressboard.
All wood strips to be of 3/4 in. wide by 3/16 in.
Pressboard strips to be of 1 1/2 in. wide by 3/16 in.
Taping of coils to be of 0.229 in.
Wire vacuum, 1.
Coil dimensions: bare, 7 5/16 in. by 1/2 in.; insulated, 7 3/8 in. by 9/16 in.
Dimensions of coils with special collars: bare, 7 5/16 in. by 5/8 in.; insulated, 7 3/8 in. by 11/16 in.

Secondary Winding

Conductor cross-section, four 0.300 in. $\times$ 0.115 in., two of double cotton covered and two of bare.
Weight, 370 lb. double cotton covered and 370 lb. bare.
60 turns in 8 S.S. coils of 15 turns each, 4 coils in series, 2 in parallel.
Insulation between turns, 1340 ft., 0.025 in. thick by 5/16 in. wide, consisting of two 0.010 in. hercules parchment and one 0.005 in. mica.
Coil dimensions: bare, 7 3/4 in. $\times$ 5/16 in.; insulated, 7 13/16 in. $\times$ 3/8 in.
Coils for special collars of 1 3/32 in.

Insulation Specification

After winding the coils they must be securely clamped to dimensions called for in the winding specifications, after which they are to have the terminals attached.

Before the coils are dipped they should have a preliminary baking for 12 hours at 250° F. (120° C.), or longer if necessary, thoroughly to dry out any moisture and shellac in the turn insulation and collars. The coils should be twice dipped when hot in 0.07-B japan and baked 12 hours at 250° F. after each dipping.

After the coils have been dipped they are to receive one pintag of 0.007 in. cotton tape for varnish treatment. The taping is to be put on according to directions given below.

Before putting on the taping the coils should be brushed over with a thin coat of 0.028 in. sticker to hold the tape to the coils. The taping should receive five brushings of 0.094 in. varnish of specific gravity 875, and should be baked after each brushing at least five hours, or until hard, at a temperature of 180° F. (85 to

95 °C.). After each taping the coils should be allowed to cool to at least 100° F. (38° C.) before the next varnishing is given.

The taping should be put on with one-half lap, except at the corners of the coils, where it should overlap not more than one-eighth at the outside edge.

With single section coils there should be added, before the taping is put on, one thickness of No. 2 cotton drill, which is to be placed over the connecting straps. The drill should extend at least 1 1/2 in. each side of the strap, and must be neatly and firmly tied down with twine and sewed together at the outside edge of the coil. A tongue of the drill should extend up the outside of the strap as far as the terminal, and should be secured to the strap with a wrapping of cotton tape.

Below is given a form of general specification as presented to purchasers of transformers. The former is for a high-voltage shell-type transformer and the latter for a high-voltage core-type transformer.

GENERAL SPECIFICATIONS FOR A RECTANGULAR "SHELL-TYPE" WATER-COOLED OIL-FILLED 70,000-VOLT 25-CYCLE SINGLE-PHASE TRANSFORMER.

General Construction.—Each transformer to consist of a set of flat primary and secondary coils, placed vertically and surrounded by a built-up steel core, the coils being spaced so as to admit of the free circulation of oil between them, which acts not only as an insulator but as a cooling medium by conveying the heat from the interior portions of the transformer to the tank by natural circulation.

The transformer to be enclosed in a boiler-iron tank, the base and cover being of cast iron. The tank to be secured to the base with a joint made oil-tight by heavy riveting and caulking.

A coil of pipe for water circulation to be placed in the oil in the upper part of the tank over the cover and surrounding the ends of the windings, the combined surface of the coil and tank being sufficient to dissipate the heat generated and thus maintain the oil and all parts of the transformer at a low temperature.

Core.—The core to be built up of steel laminations of high permeability and low hysteresis loss. The laminations also to be carefully annealed and insulated from each other to reduce eddy-current losses.

Windings.—The primary and secondary windings to be subdivided into several coils, each built up of flat conductors, wound with one turn per layer so as to form thin, high coils which will present a large radiating surface to the coil. The conductors to be cemented together with a special insulating compound, after which an exterior insulating wrapping to be applied and separately treated with an insulating varnish, making a very durable insulation.

A solid insulating diaphragm to be placed between adjacent primary and secondary coils, and to be rigidly held in position by spacing channels covering the edges of the coils.

The assembled coils, except at the ends, to be completely enclosed by sheets of solid insulation, which will interpose a substantial barrier at all points between the winding and the core.

Oil.—Each transformer to have sufficient oil completely to immerse the core, windings, and cooling coil. In order to secure the best insulating qualities and a high flashing-point, the oil to be specially refined and treated and tested for this use.

A valve for drawing off the oil to be located in the base of the tank.

Water-cooling Coil.—To consist of heavy wrought-iron lap-welded pipe with electrically welded joints, and to stand a test of at least 1000 lb. pressure per square inch.

The duty of the cooling coil is to absorb that portion of the heat that cannot be dissipated by natural radiation from the tank, which will be made to fit the transformer closely, and thus minimize the amount of oil and floor space.

Leads.—The primary and secondary leads to be brought out through the cover, and to consist of heavy insulated cables brought through porcelain bushings of ample surface and thickness.

Performance.—After a run of 24 hours at rated load, frequency, and voltage, the rise in temperature or any part of the transformer, as measured by thermometer, and the rise in temperature of the coils, as measured by the increase in resistance, not to exceed 40° C., provided the temperature of the circulating water is not greater than 25° C., and that the supply of water is normal. If the temperature of the water differs from 25° C., the observed rise in temperature should be corrected by 0.5 per cent. for each degree.

The insulation between the primary coils and the core, and

that between the primary and secondary coils, to stand a test of 140,000 volts alternating current for 1 minute, and between the secondary coils and the core a test of double the normal voltage for the same length of time.

The transformer to carry an overload of 25 per cent. for two hours without the temperature rise exceeding 55° C.

The transformer to give full kilowatt output when operating at 90 per cent. power factor without exceeding the above temperature rise.

GENERAL SPECIFICATIONS FOR A 70,000-VOLT 25-CYCLE "CORE-TYPE" OIL-COOLED SINGLE-PHASE TRANSFORMER.

Core.—The cores to be built up with laminated iron sheets of high permeability, low hysteretic loss, and not subject to appreciable magnetic deterioration. The sheets to be carefully annealed and insulated from each other in order to reduce eddy-current losses.

Windings.—The primary and secondary windings to be thoroughly insulated from each other and from the core and frame, and to stand a potential much greater than the rated voltage of the transformer.

Oil.—Each transformer to have sufficient oil to cover the core and winding when placed in the tank. The oil to be specially treated and refined in order to secure good insulating qualities and a high flashing-point.

Terminals and Connections.—The primary and secondary leads to be carefully insulated and taken from the tank through porcelain bushings, which shall have sufficient surface to prevent perceptible leakage to the frame of the transformer.

Performance.—After a run of twenty-four hours at rated load, voltage, and frequency the rise in temperature of any part of the transformer as measured by thermometer, and of the coils as measured by the increase in resistance, not to exceed 45° C., provided the temperature of the surrounding air is not greater than 25° C. and the conditions of ventilation are normal. If the temperature of the surrounding air differs from 25° C. the observed rise in temperature should be corrected by 0.5 per cent. for each degree.

Insulation between the primary winding and the core, and between the primary and secondary windings, to stand a test of 140,000 volts alternating-current for one minute, and between the secondary winding and the core a test of 10,000 volts alternating current for the same length of time.

The transformer to carry an overload of 50 per cent. for two hours without undue heating of any of the parts.

Every manufacturer has his own particular way of arranging transformer specifications. One finds it suits his purpose to give but a general description while another will sometimes present useful information and give a detailed specification covering the complete characteristics of the transformer, including a complete test record and general mechanical and electrical description.

One of the best specifications and one of great interest to all operating engineers is made up as follows:

Transformer Detailed Specification.—Form...... ×...... × ...... Tank...... Cribs...... Total I^2R......, per cent. IR ...... H.T. cable...... No......long. L. T. cable...... No......long. H.T. brushing (terminal) No......L.T. bushing (terminal) No...... Winding..... Length...... Depth .... Width....... Weight of coils and core....... Net weight of oil......Weight of tank and cribs......Total weight ......Gallons of oil...... Steel curve...... K.V.A.......Type ......Frequency...... Volts $H.T$.......Phase...... Volts $L.T$. ......Amperes......

Iron Core.—Steel...... End irons...... ×...... Iron clusters per leg...... Core irons...... ×...... Narrow iron clusters per leg...... Corners...... Weight of core...... Window.... ×...... Volume of core...... Exciting current...... Core section...... × ...... and...... square inches. Length of mag. path......Core loss, amperes..... Mag. amperes...... per cent.

H. T. Winding—Amperes...... Volts...... Turns...... Tapouts..... No. of layers.....: Sections per lag...... Insulation coil to core...... Insulation between layers...... Turns in each layer per section...... Length of section...... Space of ends...... Space between sections...... Insulation extension...... Space around core...... Size of conductor...... Diam. of conductor...... Lbs. per M. ft....... Weight...... Thickness of coil...... Mean turn...... Copper density...... Resistance (R)...... Volts per turn...... Max. volts per layer ...... I^2R...... IR......

L. T. Winding.—Volts...... Amperes...... Taps...... No. of layers...... Insulation between layers...... Section per leg...... Turns in each layer per section...... Space between sections...... Space at ends...... Diam. of conductor...... Space around core...... Length of sections...... Weight...... Insulation extension...... Size of conductor...... Lbs. per M. ft....... Mean turn ...... Thickness of coil...... (R) Resistance....... Max. volts per layer....... Copper density...... Volts per turn...... IR...... I^2R......

Like all customers' specifications these two are not complete as they do not specify the efficiency and regulation (these two important factors being on the test record sheet only); the dielectric test of the oil; the temperature of the water necessary to cool the oil; the method of supporting the transformer; lifting the transformer, nor the method employed in moving the transformer whether it be on steel-rails, a four-wheel truck or on a racket-frame.

The general characteristics of a transformer are always given in the test record sheets. A good example of such a sheet is given below and is for a high-voltage power transformer.

TEST RECORD SHEET FOR HIGH-VOLTAGE POWER TRANSFORMER (SHELL TYPE)

1000 kw., 25 cycle, 60,000-volt shell-type, single-phase transformer (water cooled).

H. T. winding 30,000 60,000 volts; L.T. winding 2300 volts.

H. T. amperes 16.66; L. T. amperes 435.

Heat Run.—

Run for 8 hours at 2300 volts L. T., and 18.5 amperes H. T. (a).

Run for 2 hours at 2300 volts L. T., and 23.2 amperes H. T. (b).

Temperatures (degrees C. at end of run).—

H. T. by resistance 35.0 for (a) and 50.0 for (b).

L. T. by resistance 37.0 for (b).

Temperature of oil 18.0 for (a) and 23.5 for (b).

Temperature of water leaving the transformer 10.0 for (a) and (b).

Temperature of frame of transformer 12.0 for (a) and 16.5 for (b).

Resistances.—

H. T. resistance at 25° C. is 1796, and for (b) is 21.32 ohms.

L. T. resistance at 25° C. is 0.0223 and for (b) is 0.02545 ohms.

TRANSFORMER SPECIFICATIONS

Insulation.—

    Voltage applied to primary and secondary and core for one minute is 120,000 volts.

    Voltage applied to L. T. and core for one minute is 5000 volts.

               (Application of alternating current).

Efficiency.—

    At 125% full-load, guaranteed eff. 98.0% and commercial eff. 98.38%.
    At 100% full-load, guaranteed eff. 97.8% and commercial eff. 98.36%.
    At 75% full-load, guaranteed eff. 97.4% and commercial eff. 98.11%.
    At 50% full-load, guaranteed eff. 96.7% and commercial eff. 96.93%.
    At 25% full-load, guaranteed eff. 93.9% and commercial eff. 94.52%.

Regulation (100 per cent. P. F.).—

    Guaranteed regulation = 1.0 per cent.

    Commercial regulation = 1.037 per cent.

General.—

    Core loss in watts = 7.335 watts.
    Excitation in amperes = 14.3.
    Impedance volts = 1.653. $((a) = 1.822$ volts.)
    Impedance watts = 9.560. $((a) = 11.720$ watts.)
    Water per minute = 2.6 gallons.
    (ingoing water 25° C.).

Detailed Specifications.—

    Height over cover = 135 in.
    Floor space = 110 in. × 59 in.
    Total weight (without oil) = 30,000 lb.
    Gallons of oil required = 1300.
    Weight of oil = 10,400 lb.
    Weight of tank and base = 7500 lb.
    Weight of large cover = 2450 lb.
    Weight of small cover = 550 lb.
    Weight of cooling coils and casing = 2500 lb.
    Dimensions of coils and casing = 70 in. × 37 in. × 32 in.
    Cooling coil (size of pipe) = 1.5 in.
    Length of cooling coil pipe = 830 ft.
    Cooling coil pipe dimensions = 30 in. × 53 in. × 96 in.
    Weight of cooling coil = 2200 lb.
    Weight of iron cover = 14,000 lb.

TEST RECORD SHEET FOR HIGH-VOLTAGE POWER TRANSFORMERS (CORE TYPE)

200-kw., 25 cycle, 57,500 volts core type, water-cooled single-phase transformer.

H. T. winding 28,750, 57,500 volts; L. T. winding 2300 volts.

H. T. amperes = 3.5; L. T. amperes = 87.

Heat Run.—

Run for 11 hours at 2300 volts L. T., and 87 amperes L. T. current (a).

Run for 2 hours at 2300 volts L. T., and 130.5 amperes L. T. current (b).

Temperatures (degrees C. at end of run).—

H. T. by resistance 30.0 for (a) and 49.5 for (b).

L. T. by resistance 38.5 for (b).

Temperature of oil 21.5 for (a) and 28.0 for (b).

Temperature at top of frame is 19.5 for (a) and 25.0 for (b).

Temperature at bottom of frame is 9.5 for (a) and 12.5 for (b).

Room temperature for (a) is 18.5 and for (b) 17.5° C.

Resistance.—

H. T. resistance at 25° C. is 56.6 ohms on winding connected for 28,750 volts.

H. T. resistance for (b) is 68.1 ohms.

L. T. resistance for 25° C. is 0.1168 ohms, and for (b) is 0.1307 ohms.

Insulation.—

H. T. to L. T. and core 115,000 volts alternating current for one minute.

L. T. and core 10,000 volts alternating current for one minute.

Efficiency.—

At 100% load guaranteed eff. 96.7% and commercial eff. 97.9%.
At 75% load guaranteed eff. 96.2% and commercial eff. 97.8%.
At 50% load guaranteed eff. 95.0% and commercial eff. 96.8%.
At 25% load guaranteed eff. 91.2% and commercial eff. 94.5%.

Regulation (100 per cent. P. F. = 1.6 per cent.).—

General.

Core loss in watts = 2.865.

Excitation in amperes = 10.3.

Impedance volts (57,500 volt winding) = 1.017.

Impedance watts = 1.750.

Detailed Specifications.

Height over all = 103 in.
Floor space = 47 in. ×64 in.
Total net weight (without oil) = 10,000 lb.
Weight of oil = 3,000 lb.
Gallons of oil required = 150.
H. T. voltage taps = 57,500, 55,000, 47,500 and 45,000 volts.

Up to the present time natural-cooled oil-insulated single-phase transformers have been built in sizes of 3000 kv-a, and air-cooled transformers up to 4500 kv-a at 33,000 volts.

Comparing the various types of European and American made transformers in so far as their over-all dimensions, kilowatts per square foot, kilowatts per cubic foot, cubic foot of air per minute, gallons of oil for a given size, etc., we find there exists quite a difference. For instance, taking two of the largest sizes made in the respective countries, we have, for the air-cooled type:

4000-KW., 33,000-VOLT, AIR-COOLED SINGLE-PHASE TRANSFORMER

(American manufacture)
(Dimensions, 90 in. ×74 in. ×137 in. high)

Kilowatts per square foot	86.5
Kilowatts per cubic foot	7.6
Cubic feet of air per minute	6.750
Frequency	25

1700-KV-A., 6600-VOLT, AIR-COOLED (OIL INSULATED) THREE-PHASE TRANSFORMER

(European manufacture)
(Dimensions, 86 in. ×25 in. ×98 in. high)

Kilowatts per square foot	55
Kilowatts per cubic foot	6.7
Gallons of oil	640
Gallons per kilowatt	0.376
Cubic feet of air per minute	7.500
Frequency	25

The European type referred to here is oil-insulated and cooled by means of an air blast at the outside of the tank. The American type referred to here is commonly known as the air-blast transformer and is cooled by forced air circulation through core and coils.

Also, comparing the largest water-cooled oil-insulated transformer made in Europe with an ordinary 4000 kw. standard American design of the same type, we have:

4000-KW., 33,000-VOLT, OIL-INSULATED WATER-COOLED SINGLE-PHASE TRANSFORMER

(American manufacture)
Dimensions, 107 in. ×63 in. ×150 in. high)

Kilowatts per square foot	86
Kilowatts per cubic foot	6.85
Gallons of oil	1950
Gallons per kilowatt	0.49
Full-load efficiency	98.9 per cent.
Half-load efficiency	98.5 per cent.
Gallons of water per minute	19.5
Frequency	25

5250-KV-A., 45,000-VOLTS OIL-INSULATED, WATER-COOLED, THREE-PHASE TRANSFORMER

(European manufacture)
(Dimensions, 82 in. ×59 in. ×136 in. high)

Kilowatts per square foot	162
Kilowatts per cubic foot	14.2
Gallons of oil	1130
Gallons per kilowatt	0.215
Efficiency at full-load	98.95 per cent.
Efficiency at 50 per cent. load	98.85 per cent.
Regulation (P. F. =1.0)	0.62 per cent.
Regulation (P. F. =0.8)	3.0 per cent.
Frequency	50

Transformers of a much greater size than 5,000 kv-a at 45,000 volts have not yet been built in Europe. In American single-phase units of 7500 kw., and three-phase units of 14,000 kw. are in operation at voltages above 100,000 volts.

The three following high-voltage power transformers will serve to show what is being done in this direction:

7500-KW., 60,000-VOLT, OIL-INSULATED, FORCED OIL-COOLED, SHELL-TYPE THREE-PHASE TRANSFORMER

Kilowatts per square foot	80
Kilowatts per cubic foot	5.65
Gallons of oil	4000
Gallons of oil per kilowatt	0.53
Efficiency at full-load	98.95 per cent.
Weight without oil	42.5 tons.
Frequency	25

5750-KW., 138,500-VOLT, OIL-INSULATED, WATER-COOLED, SHELL-TYPE SINGLE-PHASE TRANSFORMER

Kilowatts per square foot	71
Kilowatts per cubic foot	5.3
Gallons of oil	2500
Gallons of oil per kilowatt	0.7
Efficiency at full-load	98.8 per cent.
Efficiency at 75 per cent. load	98.7 per cent.
Efficiency at 50 per cent. load	98.3 per cent.
Efficiency at 25 per cent. load	96.9 per cent.
Total weight in tons	28
Frequency	60
Test voltage	280,000 volts.

(Length of high-voltage winding is 4 miles)

10,000-KW., 100,000-VOLT, OIL-INSULATED, FORCED OIL-COOLED, SHELL-TYPE THREE-PHASE TRANSFORMER

Kilowatts per square foot	58.7
Kilowatts per cubic foot	3.25
Gallons of oil	1500
Gallons of oil per kilowatt	0.75
Efficiency at full-load	99 per cent.
Weight without oil	60 tons.
Frequency	60

(The primary winding (high-voltage winding) consists of 10 miles of copper conductor)

Practically every high-voltage alternating-current three-phase system operating at 55,000 volts and above are given in the table "Modern High-voltage Power Transformers operating at the Present Time." The transformer connections are as 38 for star against 21 for delta (the star connections being in some cases solid grounded and in others grounded through a high resistance —while the delta connections are all insulated). In Chapter IV several important advantages and disadvantages are given for delta and star connections, in looking through the list of the highest voltage transformer installations practice appears to point toward the star connection.

APPENDIX

*MODERN HIGH VOLTAGE POWER TRANSFORMERS OPERATING AT THE PRESENT TIME.

Operating systems	Transformed kilowatt capacity	Connection of system	†Phase ‡	System frequency	Transformer voltage of system
1. Sierra and San Francisco	3,750	Star	1	60	140,000
2. Power Company	2,233	Star	1	60	104,000
3. Eastern Michigan Power Co	3,000	Delta	1	60	140,000
4. Hydro-Electric Power Comm-n	1,250	Delta	1	25	110,000
Hydro-Electric Power Comm-n	750	Delta	1	25	110,000
5. Mexican Northern Power Co	2,500	Star	1	60	110,000
6. Great Western Power Company	10,000	Delta	3	60	110,000
Great Western Power Company	5,000	Delta	1	60	90,000
7. Mississippi River Power Co	9,000	Star	3	25	110,000
8. Grand Rapids Mich. Power Co	3,750	Delta	3	60	110,000
9. Georgia Power Company	3,333	Star	1	60	110,000
10. Truckee River G. E. Company	1,000	Star	1	60	104,000
11. Yadkin River Power Company	6,250	Star	3	60	104,000
12. Great Falls Water P. & T. Co	2,400	Delta	1	60	102,000
13. Southern Power Company	3,000	Delta	1	60	101,200
14. Central Colorado Power Co	3,333	Star	1	60	100,000
15. Tata Hydro-Electric Power Co	3,333	Star	1	60	100,000
16. No. State Hydro-Electric Co	2,750	Star	1	60	100,000
17. Yadkin River Power Company	2,500	Star	1	60	100,000
18. Shawinigan Power Company	14,000	Delta	3	60	100,000
Shawinigan Power Company	12,000	Delta	3	60	85,000
19. Central Colorado P. Company	2,500	Delta	1	60	90,000
20. Rio de Janeiro L. & P. Co	2,000	Star	1	60	88,000
21. Appalachian Power Company	6,000	Delta	3	60	88,000
Appalachian Power Company	1,500	Delta	1	60	85,000
22. Saõ Paulo Electric Company	1,500	Star	1	60	88,000
23. Toronto Power Company	6,000	Star	3	25	86,000
24. Tata Hydro-Electric Company	3,120	Star	1	60	85,000
25. Mexican L. & P. Co. Ltd	6,000	Star	1	60	85,000
Mexican L. & P. Co. Ltd	5,600	Star	1	60	81,000
26. Madison Riven Power Company	900	Delta	1	60	80,000
27. Utah Light & Power Company	333	Delta	1	60	80,000
28. Butte Elect. L. & P. Company	333	Delta	1	60	80,000
29. Telluride Power Company	1,500	Star	1	60	80,000
30. Kataura-Gawa Hydro-Elect. Co	3,500	Star	1	50	77,000
31. Southern Calif. Edison Co	1,667	Star	1	50	75,000
32. Pennsylvania W. & Power Co	10,000	Star	3	25	70,000

*The connection of systems given here is not strictly correct as changes from delta to star and from star to delta are constantly being made to suit new conditions.

†This denotes single or three-phase transformer—the systems themselves in all the above cases being three-phase.

APPENDIX

MODERN HIGH VOLTAGE POWER TRANSFORMERS OPERATING AT THE PRESENT TIME.—(Continued.)

Operating systems	Transformed kilowatt capacity	Connection of system	Phase θ	System frequency	Transformer voltage of system
33. Missouri River Power Co.	1,500	Star	1	60	70,000
34. Southern Wisconsin Power Co.	1,000	Delta	1	60	70,000
35. Kern River Power Company	750	Star	1	50	67,500
36. Northern Calif. Power Co.	4,000	Star	3	60	66,000
37. Yakima Valley Power Company	1,000	Star	1	60	66,000
38. Central Georgia Power Co.	2,500	Star	1	60	66,000
39. Northern Hydro-Electric Co.	1,100	Star	1	25	66,000
40. Eastern Tennessee Power Co.	1,100	Star	1	60	66,000
41. Idaho-Oregon L. & P. Company	750	Delta	1	60	66,000
42. Spokane & Inland E. R. Co	1,500	Star	1	25&60	66,000
43. Nagoya Electric Power Co.	1,000	Star	1	60	60,000
44. Washington Water Power Co.	2,200	Star	1	60	60,000
45. Mexican L. & P. Co. Ltd.	2,000	Star	1	60	60,000
46. East-Creek L. & P. Company	2,800	Delta	3	25	60,000
47. Great Northern Power Co.	3,000	Delta	3	25	60,000
48. Niagara-Falls Power Company	1,750	Delta	1	60	60,000
49. Pacific Coast & Electric Co.	1,750	Star	1	60	60,000
50. Guanajuato Power & E. Co.	1,000	Delta	1	60	60,000
51. Jhelum River Electric Co.	1,000	Delta	1	25	60,000
52. Michiacan Power Company	1,300	Star	1	60	60,000
53. Elect. Development Company	2,670	Delta	1	25	60,000
54. Puget-Sound Power Company	2,333	Star	1	60	60,000
55. Canadian-Niagara Power Co.	1,250	Star	1	25	57,000
56. Portland L. & P. Company	3,367	Star	1	60	57,000
57. Calif. Gas & Electric Co.	840	Star	1	60	55,000
58. Pacific Coast Power Company	3,333	Star	1	60	55,000
59. Winnipeg L. & P. Company	1,800	Delta	1	60	55,000
60.					

INDEX

Ability of a transformer to deliver current at a constant voltage, 16
Admittance, 31, 32
Advantage of the delta-delta to delta-delta, 74, 75
    delta-star to star-delta system, 74, 75
Advantages of iron and air reactance coils, 159
Aging of the iron, 11, 15
Air and iron reactance coils, 159
Air-blast transformers, 7, 120, 144
Air-chambers for transformers, 144
Air-cooling of transformers, 120-122
All-day efficiency, 14, 15
Alloy steel in transformers, 11, 246
Alternating currents dangerous and impracticable, 3
    fought in the Law Courts, 3
American and European engineers, 79
American I. E. E., 239
Amount of heat developed in a transformer, 6
    hysteresis in a given steel, 8
Anhydrous copper sulphate, 132
Arc-series lighting, 178-185
Assembly of large power transformers, 136-148
Auto-transformers and transformation, 33, 170-177
Automatic regulators, 224-226

Balancing transformer, 26, 27
Best shape of transformer core, 11
Best operated system, 74
Blotting paper, 133
Breathing action, 125
Building of fire-proof construction, 8

Capacity in kw., 3, 52
Causes of transformer failure, 145, 154

Centralized system over 200,000 kw., 3
Central station engineers and managers, 153
Cheap copper space and alloyed iron, 153
Cheapest cost of system, 74
Changing of frequency (using transformers), 83
Change any polyphase system into any other system, 35
Chief danger of fire, 8
Choice of transformer system connections, 70, 74
Cleaning of transformer cooling-coils, 125
Close regulation, 156
Coefficients of resistivity in copper —Table, 240
Comparison of shell and core type transformers, 150-153
Comparison of single-phase and polyphase transformers, 43
Comparative weights of transformers, 42
Commercial manufacture of 175,000 volt power transformers, 3
Common return wire, 32, 39, 49
Combination method of cooling transformers, 122
Compensators, 218-220
Connections for grounding three-phase transformers, 81
Connections for grounding two-phase transformers, 39
Consolidation and concentration of systems, 158
Constant current transformers, 14, 178, 185
    potential transformers, 4, etc.
Construction of large power transformers, 136 144
    of constant current transformers, 182-184

INDEX

Contact-making voltmeter, 224-226
Control of the designer, 9
Conversion of one polyphase system into another, 20
Conventional connections for transformers, 56
Cooling of transformers, 6, 117-130
Cooling-coil cleansing, 125
Cooling medium, 153
Copper or $I^2 R$ loss in transformers, 9, 16, 250-252
Cost of a given volume and area, 9, 11
Core of the transformer for a given service, 5, 75, 153
    loss or iron loss, 8, 11, 14, 245-248
    type transformers, 10, 24, 125, 136, 137-139, 150-153, 158, 262
Cost of total losses, 5
Current, short-circuit, 154, 156, 257
    transformers, 186-208

Dangerous and impracticable current, 3
'Dead," short-circuit, 154
Dear copper space and ordinary iron, 153
Delta-connected systems, 42, 53
    -star systems, 47, 53, 73
    -star merits, 53, 73
    -delta systems, 47, 54, 73
    -delta merits, 48, 53, 73
Demerits of the delta-delta system, 74-75
Design of oil-insulated transformers, 118
Determination of temperature rise, 7
Development of transformers, 2
    of the art of transformer construction, 1
Diametrical system, 116
Difference of potential, 34
    of opinion as regards grounding, 28
Difficulties of three-phase operation, 84-96
Different ways of applying transformers, 22

Disadvantage of the delta-delta system, 74, 75
    of the "T" (two transformer) system, 80
Disc-shape coils, 129
Distribution transformers, 28
Disturbed system due to incoming surges, 75
Dividing each secondary coil of transformers, 24
Double-delta system, 109, 113, 116
    -star system, 109, 114, 116
    -tee, system, 114, 116
Drop due to load, 31, 32
Drying of insulation, 145
    out of power transformers, 146-148
Duration of tests, 13

Early development of transformers, 1
Earth connections, 163
Earthing of the neutral of transformers, 161
Economy of copper, 30
Eddy-current loss, 8, 245-247
Edison, Thos. A., three-wire system, 51
Effect of various iron substances, 11
Efficiency of transformers, 14, 16, 153, 252, 253
Electrical characteristics of a transformer, 5
Electro-magnetic properties of alloy-steel, 11
Electro-static capacity of parts too high, 95
Equal output and change in copper loss, 9
Engines, 3
Exciting current, 9, 10
External and internal choke-coils, 159

Factory, transformers from the, 26
Faraday's historic experiments, 1
Favor of the shell-type transformer, 152
    of the core-type transformer, 152

INDEX 275

Feeder-regulators, 209-220
Ferranti's S. Z. de., modification of Varley's method, 1
Filter-paper, 125
Fire-risks of air-blast transformers, 7
Five-wire two-phase system, 34
Fluid of some impregnating compounds, 12
Fluxes (magnetic) in transformers, 42
Forced current of air-cooling, 120
current of water-cooling, 122
oil circulation, 127
Form of transformer coil, 9
Four-wire two-phase system, 30, 33
Freezing of water, 124
Frequency changer (using transformers), 83
Fundamental equations in design, 13
principals, 3
Fused-switches, 169

General construction, 260
Generators, 3
Generating and receiving stations, 144
Governing factors of systems, 74
Grading of insulation on the end turns, 151
Graphical methods, 17, 23, 40, 43, 44, 46, 53, 111
Greater coal consumption due to exciting current, 9
generating station equipment, 10
Ground connections, 163-165
Grounded systems, 28, 37, 29, 74, 81-82, 84, 161
Grounding of the neutral point, 39, 74, 76, 81, 161

Heat test of transformers, 232, 233
High frequency currents, 159
voltage operation of transformers, 2
stresses, 76, 79, 86
surges, 74, 75
transformer troubles, 94
installations, 125, 136-144
specifications, 258, 267

History of alternating currents, 3
of delta and star systems, 3
of 50,000 and 60,000 volt systems, 3
Hourly temperature readings, 153
Hydrochloric acid-cleaning of coils, 125
Hysteresis, 8, 10, 245-248
loss, 8

I R drop, 18, 19, 67, 91
I^2 R loss, 9, 16, 20
Incoming surges, 75
Increased hysteresis and higher temperatures, 9
temperatures due to excessive iron losses, 9
Inductive load on transformers, 10, 17
reactance (X_s), 18
Inherent reactance, 154
Impedance (Z), 19, 24, 31, 67, 91, 250-252
Important advantages of the delta-star to star-delta, 74, 75
factors in the make-up of a transformer, 8
Impregnating of transformer coils, 11
Improper drying out of transformers, 95
Impulses from transmission line, 75
Installation of large power transformers, 144-148
Instantaneous values of the currents, 40, 47
Insulation, 11, 117, 227, 228
specification, 259
Internal and external choke-coils, 159
Iron and air reactance coils, 159
or core loss, 8, 9, 11, 14, 245-248

Joints of ground wire, 162, 164

Kirchhoff's law, 44
Kv-a capacity of transformers, 3, 52
K-w capacity of transformers, 69, 74, 80

276   INDEX

Lamps, arc and incandescent, 178–185
Large commercial power transformers, generators, etc., 3
Law courts and alternating currents, 3
Leakage reactance of the transformer, 154
Least potential strain on system, 74
Liberal oil ducts between the various parts, 6
Lighting transformers, 31
Limiting feature of transmission, 2
   resistance, 163
Limit of temperature rise in transformers, 6
Line-drop compensator, 221–225
Long-distance transmission lines, 151
Losses, 8, 79
Loss due to magnetic leakage, 10
   to magnetizing current, 10
   in revenue due to transformer failure, 6
Low average operating temperature, 6
Lowering the frequency of supply, 9, 83
Lowest price for transformers, 74

Magnetic densities of transformers, 14
   leakage, 16
Magnetizing currents in transformers, 11
Mechanical failure of transformers, 76, 152, 153, 155, 160
Mercury thermometers for temperature readings, 147
Merits of the delta-star to star-delta system, 74, 75
   of three-phase units, 43
Method of connecting three-phase systems, 42
   two-phase systems, 30, 33
   of insulating transformers, 11
Methods of cooling transformers, 117–130
Meyer and Steinmetz systems the same thing, 106 (Fig. 101 A)

Modern high voltage power transformers, 270, 271
Moving of coils and core, 94, 154, 160
Motor auto-starters, 219, 220

National Board of Fire Underwriters, 227
National Electrical Code, 28
Negative direction, 40, 56, 244, 245
Neutral point, 39, 49, 76, 81, 86, 161

Ohm's Law, 248, 252
Ohmic drop, 16, 42
Oil for transformers, 127, 130–135
   -filled self-cooled transformers, 7
   transformers in separate compartments, 8, 144
Open-delta "V" system, 51, 52, 53, 72
Operation of transformers, 153–169, 178–184
Operating at the end of long lines, 11
   engineer's difficulties, 64
Operation of constant current transformers, 178–185
Opposition-test, 236
Over-potential test, 229

Parallel operation, 22, 38, 56, 64, 66, 78, 91, 239
Phase relations, 22, 59, 68, 74, 87
Phase-splitting methods, 21
Points in the selection of transformers, 5
Polarity of transformers, 55, 58, 59–64, 243–245
Polyphase regulators, 220–226
Positive direction, 40, 56, 244–245
Power (P) or electrical energy, 31
   factor, 10, 16, 19, 22, 203–204
   limiting capabilities of reactance, 158
   transformers, 43, 56, 66, 123, 145, 153, 267–271
Practicable 50,000 and 60,000 volt systems, 3
Primary windings of transformers, with R and X_s, 5

INDEX

Principal. argument for grounding transformers, 28
Puncturing of the insulation between turns, 94, 117

Raising current, 205
Ratio of high voltage to low voltage turns, 65, 67
   of iron and copper loss, 5
   of transformation, 65, 67, 239–241
Reactance, internal and external, 154, 160
Recognized commercial operation of 60,000 volts, 3
Receiving station transformers, 74
Record sheet of tests, 264–269
Regulation, 10, 16, 22, 155, 223, 253–256
Regulators and compensators, 209, 226
Relays for protection of transformers, 165–167
Reliability of transformers in service, 80, 151, 154
Resistance (R), 18, 41, 248–250
   per phase, 31
   of wiring, 31
Revenue effected imperfect iron of a low grade, 11
   loss due to transformer failure, 6
Roasting of oils due to over-load, 154

Safety to life and property, 5, 154
Salt for the earth connections, 165
Scott, Chas. F., system, 53, 80, 98, 106
Secure the desired copper loss, 9
Self-cooling transformers, 117–120, 144
   induction, 18
Short-circuits, 28
   -circuit stresses, 154, 156, 257
Shell type transformer, 2, 10, 73, 79, 128, 139–144, 260
Series enclosed arc lighting, 178–180
   lighting from constant current transformers, 205

Series-parallel operation, 152
   step-up transformer, 205
   transformers, 186–208
Simple transformer manipulations, 22
Single-phase systems, 22, 42, 83
   from three-phase, 82, 83
Six-phase transformation and operation, 109–116
Solid compounds for impregnation of windings, 11
   insulation, 11
Spark-gap—Table, 228
Specifications of transformers, 258–270
Star or "Y" system, 41–53
   -delta system, 47, 53, 73, 74
   merits, 47, 73, 74
   -star system, 47, 53, 55
   versus delta, 71, 73, 74
Station equipment and transmission lines greater, 10
Standard rules of the A. I. E. E., 239
   voltage test, 231
Steam turbines of 40,000 h.p. capacity, 3
Steinmetz, Chas. P., system, 100, 106
Step-up current transformers, 205
   -down transformers, 4
   -up transformers, 4, 23
Strain between high and low voltage windings, 27
Stillwell, Lewis B., regulator, 210
Surges, 75, 159, 163
System giving cheapest cost, 74
Switches, 76, 165
Switching and its advantages, 76, 165, 109–116

Taylor, William T., system, 37, 71, 80, 106, 255
"T" two-transformer system, 37, 53, 71, 80, 98
   three-transformer system, 37, 71, 80
Telephone transformers, 206
Temperature rise of the oil, 128
   of electrical apparatus, 7, 147

Temperatures, 123, 125, 128, 152, 232-239
Terminals and connections, 262
Test record sheets, 264-267
Testing cooling coils for transformers, 146
Tests specified by the National Board of Fire Underwriters, 12
    of copper loss and impedence, 250-252
    of efficiency, 252, 253
    of iron or core loss, 245-252
    of insulation, 13, 227-232
    of polarity, 243-245
    of ratio of transformation, 239-243
    of regulation, 253-256
    of resistance and I R., 248-250
    of short-circuits, 257
    of temperature, 232-239
Third harmonics, 75
Three-phase systems, 22, 39, 53
    to single-phase, 82, 104, 105
    to six-phase, 109-116
    to three-phase, two-phase,
    to two-phase, 53, 97-109
    two-phase methods, 107, 109
    group of 18,000 kw., 3
Three-wire service transformers, 24, 26
    two-phase systems, 30, 33
Time-limit relays, 169
    -phase, 40
To secure the desired copper loss, 9
Transmission engineers, 153
Transformer development, 1
    of 14,000 kw capacity, 3
    regulation, 10, 16, 17
Transformers of identical characteristics, 22
    in separate compartments, 8
Troubles experienced with high voltage transformers, 94
Turbines, 3
Turbo-generators of 30,000 kw., 3
Two-phase systems, 22, 30, 33
    to three-phase, 22, 53
    three, four and five wire, 34

Two-phase, to single-phase, 35, 36
    multi-wire distribution, 36
    parallel combinations, 38
    three-phase methods, 107, 109

Unbalancing, 32, 52, 68, 75, 85
Unduly moist transformers, 145
Unlimited power behind transformers, 156
Use of reactance, internal and external, 154
    of resistance in the neutral of grounded systems, 154
Useful energy delivered, 9

Vacuum process of impregnation, 11
Varley's method, 1
"V" or open-delta system, 42, 51, 53, 72
Variation of core loss, 248
    of copper loss, 251
Vector representation, 17, 30, 33, 40, 44, 46, 53, 116
Viscosity, 130, 132
Voltage compensator, 221-224
Voltmeter—contact-making, 224, 225

Water available and not expensive, 7
    -cooled transformers, 7, 122, 144, 258, 261
    -freezing difficulties, 124
    -wheels, 3
Wattmeter, 203, 204, 207
Watts radiated per square inch of surface, 119
Weak link in the insulation, 152
    mechanically, transformers, 154
Weight, comparative, 42, 79
Windings, specifications, 261, 262

X_s or inductive reactance, 18

Y or admittance, 31
"Y" or star connection, 41
    three-phase to two-phase system, 98

"Z" (series transformer) connection, 200, 201

www.ingramcontent.com/pod-product-compliance
Lightning Source LLC
LaVergne TN
LVHW012229010225
802757LV00008B/618